Project Management Blunders

Lessons From the Project That Built, Launched, and Sank *Titanic*

By Mark Kozak-Holland

First Edition

Multi-Media
P u b l i c a t i o n s Inc.

Oshawa, Ontario

Project Management Blunders: Lessons From the Project That Built, Launched, and Sank *Titanic*

by Mark Kozak-Holland

Managing Editor:	Kevin Aguanno
Copy Editor:	Susan Andres
Typesetting:	Peggy LeTrent
Cover Design:	Kevin Aguanno
eBook Conversion:	Charles Sin

Published by:
Multi-Media Publications Inc.
Box 58043, Rosslynn RPO, Oshawa, Ontario, Canada, L1J 8L6.

http://www.mmpubs.com/

Paperback	ISBN-13: 9781554891221
eBook Version	ISBN-13: 9781554891238

Also available in various ebook formats.

This book replaces *Titanic Lessons for IT Projects* (ISBN 9781895186260) published 21 September 2005.

Published in Canada. Printed simultaneously in the U.S.A, the United Kingdom, and Australia.

CIP data available from the publisher.

To my wife Sharon and children
Nicholas, Jamie, and Evie.

Table of Contents

Acknowledgments

Variations of this book have been available since 2002 when it came out. The initial idea occurred to me in 1994, but the book took many years to complete. Since that time, I have spoken to many people who helped shape the book. I would like to start by mentioning some team members involved in the project who brought forth the idea, namely, Denis Cormier, Angelo D'Amore, and Andre Beauchamp. I would also like to thank Sheila Richardson and Jill Batistick for helping transform the first brief into a workable manuscript and Tara Woodman and Merrikay Lee for helping transform it into the book's first copy. I would also like to thank Alister McGuinness and Glenn LeClair for their comments, and Andrew Michael (Maritime Engineer) for his detailed reviews and input on the technical aspects of the Olympic-class project.

I am indebted to many Project Management Institute chapters and the International Project Management Association (IPMA) for sponsoring the "Projects Lessons from *Titanic*" speaking event, which led to many conversations with project managers who brought forth many comments and ideas about their "*Titanic* projects."

I would like to thank Kevin Aguanno for helping pull this book together and for driving the initiative. I am indebted to my wife and family who have been so gracious in allowing me to continue this writing project at the expense of our valuable time together.

If you have any ideas for improving this book, please contact me by e-mail. Your feedback can be incorporated into a future edition.

Mark Kozak-Holland

E-mail: mark.kozak-holl@sympatico.ca

Preface

This book is primarily about project management. It takes one of the most renowned events ever and extracts lessons from it to turn project management theory into reality. So, why the *Titanic* case study? Most people know the *Titanic* story from the intense media focus through documentaries on television, the salvage operation, the touring *Titanic* exhibition, and the 1997 James Cameron movie. Typically, these focus on the last two days of the voyage and the last hours of the disaster. But what about the four-year project that delivered the Olympic-class ships? Was it significant? What impact did it have on the disaster? What can we learn from the project about project management? What can we learn and transfer to today's projects?

The book examines the *Titanic* project foremost from a project management perspective and through the modern lens, as defined by the Project Management Body of Knowledge (PMBoK®)—very different, for example, to *Titanic—The Ship Magnificent*, which examines the project completely from an engineering perspective. This book expands on the project and the phases, but most important, examines it through the nine knowledge areas of PMBoK®, very much how a modern project manager would approach the project.

The project to build the Olympic-class ships is a significant case study in project management history. It was the largest project at its time, 1907–1912, and nothing surpassed it in scale, as it delivered the largest moving objects ever built. It embraced the latest in emerging technologies, and it was delivered by the premier shipbuilder of the time with the highest possible reputation, using the day's best practices for people, process, and technology (equipment).

Yet, the project is also a case study in poor project management leading to a catastrophic project failure (or disaster). Root causes included meddling stakeholders, different and competing agendas, compromises, a rushed job, more compromises, changes late in the project, improper testing, and a poor implementation, where the biggest learning lessons are for projects today.

Why is this important? Most organizations have suffered through their "*Titanic* project," a project that results in a catastrophic failure or disaster. Most project managers will admit living through at least one of these in their career, and from my experience, I have seen my share of these, which probably gave me the most important learning experiences I have had. Usually, the failure's roots, as in the case study, occur in the project itself.

A project that fails on or about implementation can be expensive. But the most expensive and critical failures are those that happen weeks or months post-implementation, after the project has been deemed completed. These are unpredictable, unexpected, and by far, the costliest. In my encounters with this type of failure, typically, the project team has been disbanded and allocated to new projects, long before the problems emerge. In addition, the operations group grapples with an unknown solution as they have been brought much too late into the project to appreciate and understand the solution well enough to manage it adequately.

One challenge in investigating failures weeks or months into post-implementation is convincing senior-level executives that the failures' root is poor decision making in the project itself. Often, decision making was hindered by poor investments in the project and a focus on functional, rather than nonfunctional, requirements—what the solution does rather than how it does it. The project team drifts through the project and allows these poor decisions to pass without making any noise or outright challenges. Often, the bombshell drops that something must be done, and it comes far too late to stem the disaster. I hope that reading this book gives you enough insight and best practices to step back and prevent the failure in time.

Author's Perspective on the Book

For many years as a consultant and project manager, I was involved in designing, creating, delivering, and running solutions. Many of them were continuously available solutions built to withstand any interruptions to service. I was also involved in completing project postmortems and investigating the significant solution failures in operation. One chief challenge in investigating these types of failures was convincing management that the root causes were in the project itself, preceding the implementation.

As I began to understand these root causes more deeply, I started to draw parallels from some notable disasters of the twentieth century: the *Titanic*, the *Hindenburg*, Three Mile Island, Chernobyl, and the space shuttles *Challenger* and *Columbia*. As a result, I investigated these disasters, and I researched the *Titanic* extensively. The *Titanic* kept coming to the forefront of my thinking as the volume of research material on the disaster is astounding. The *Titanic* provided an excellent analogy for why solutions fail catastrophically when put into operation literally four days into production. These four days of operations were chaotic and littered with mistakes—the breakdown in early warning systems, the failure to manage incoming intelligence, and succumbing to business pressures to better the *Olympic*'s best Atlantic crossing time. Once the collision with ice occurred, the compromises in the project's design and poor decision making doomed the *Titanic*. Most important, the root causes lay in the four-year project itself and the decision making that went on. The case study provided an opportunity to explore the gray area between project implementation and operations, which raised the questions of when a project should end and when it can be deemed a success or failure.

My first book with Multi-Media Publications, *Avoiding Project Disasters* (2005), was created to help readers better understand how problems in operation or production correlate back with the starting project, as a result, culminating in a better understanding of risks and decision making and helping in the better running of projects. Since the book's publication, I have had many conversations with readers, mostly project managers. The overwhelming response I had on the book indicated the need for a more high-level, shorter, less technical version that captures the essence of the *Titanic* story, which proved the central feature. Hence, the second book *Titanic Lessons for IT Projects* (2007).

The book you are now reading is a further evolution in the examination of the same case study but from a deeper project management perspective, which also reflects my career growth in project management. The book is based on further feedback from the project management community and the latest research on the case studies brought forward by the *Titanic* community, such as the grounding theory, further discoveries from the wreck site, and the ever-increasing body of knowledge of the subject. It squarely focuses on project management, and the most important feature is it allows the reader to explore complex project problems not covered in the PMBoK®, for example, the relationship with the executive sponsor and project manager and keeping the sponsor engaged through the project. It also helps the reader prepare for sitting the Project Management Professional (PMP) or the IPMA Level-D examination and project management lessons beyond.

How to Use This Book

Each chapter is structured to follow a phase of the project. Lessons learned are presented at the end of each chapter, as these should typically be extracted throughout the project and not just at the project's conclusion. The final section of each chapter has been created for educators, and it provides selected discussion points.

CHAPTER 1

Project Initiation

In This Time Frame
○ Early 1907—Initial idea for White Star to deliver a new fleet of liners into service.

Overview

This chapter looks at why and how White Star began a project to build a new fleet of liners, established the business need, and shepherded the project through the first-phase gate. Today, the start of any project is known as **initiation**, which sets up the project through a **project charter**.

Background to the Project

To understand better the causes to the *Titanic*'s disaster, we must go back to 1907 and review the business situation facing White Star, the company to build the fateful ship. The company had just come through a crippling rate war with its competitors. Worse still, the principal competitor Cunard liners had just received a new fleet of ships, including the much larger and luxurious *Lusitania* and *Mauretania*. These had been built with a British government subsidy, as the government could use the ships in wartime. These ships were built for speed, using the latest in steam turbine technology, and they held the Blue Riband for the fastest transatlantic crossing. For White Star, there was a pressing **business need** for something to be done. White Star's aging fleet of liners was grossly inadequate to compete with this stiff growing competition from both Cunard Liners and North German Lloyd.

Global Economic Climate

In 1902, there were two superpowers in the world, the United Kingdom and the US, both achieving their economic positions on the backs of two industrial revolutions. The first industrial revolution (1750–1890), based on coal, steam, and iron, powered the United Kingdom to the first global superpower. Toward the end of the nineteenth century, the US started to catch up. The second industrial revolution (1890–1940 and based on the internal combustion engine and electricity) had a more profound impact in the US than in the United Kingdom. A transportation and communication revolution was sparked because of both of these. Not surprisingly, all these critical emerging technologies (steel ships, steam turbines, electric power, wireless telegraphy, telephone communications, and automated control systems) from the industrial revolutions play an important role in this story.

J. P. Morgan Takes Over White Star and Sets a Direction

Between 1900 and 1902, transatlantic passengers increased from 875,000 to 1.149 million, a 24% rise. J. P. Morgan, the richest man in the world, had taken over White Star in 1902 and put it under International Mercantile Marine (IMM), the conglomerate of shipping companies. Morgan was in the transportation business, and he had built a railroad empire, so moving into shipping was a natural progression. The US was behind the United Kingdom in this marketplace, and it didn't have the latest in emerging technologies. IMM allowed White Star to continue to use its name, fly the British Ensign, and man the ships with British crews to avoid the higher US port taxes. So, this seemed a British company rather a US one. Morgan expected a return on his investment, but not right away.

Bruce Ismay Lays Out a Vision

Bruce Ismay had been the chairman of White Star since he took over his father's position in 1899. In 1904, Morgan gave Bruce Ismay complete control of IMM and made him the president. Bruce Ismay had recognized the subtle changes in the liner market and the emerging "tourist class." He had a **vision** for a new fleet and needed input in how it could be realized. He met with Lord Pirrie, the chairman of Harland and Wolff (see Figure 1.1.), the preferred shipbuilders for White Star. Bruce Ismay had continued his father's close relationship with Harland and Wolff. The two companies had an agreement where all White Star ships were built at the Harland and Wolff shipyard, provided Harland and Wolff did not construct ships for the competition. White Star was Harland and Wolff's largest Customer.

Figure 1.1. Harland and Wolff was White Star's preferred supplier with a long and productive business relationship spanning many decades.[1]

Pirrie the Technologist

Harland and Wolff was probably the premier shipbuilder in the world with a renowned reputation for quality. Pirrie, selected as a trusted advisor, was on the company board of White Star. He was a technologist and expert on changes in new emerging technologies affecting the world of ships. He

15

was a progressive shipbuilder and always on the lookout for new ideas.[2] The turn of the twentieth century saw many rapid advances in new patents and inventions from all scientific and engineering disciplines. These were more readily incorporated into shipbuilding than elsewhere, because of the fiercely competitive environment. The rate of technology change was so fast that the latest fittings on new ships were outdated within several years. Harland and Wolff knew of this constant technological evolution, and they successfully combined this with the best traditional practices.

Bruce Ismay and Pirrie Create a Strategy to meet the Vision

In 1907, over dinner at Pirrie's London home, both men (see Figure 1.2) discussed the changing business landscape and the competition's hopelessly outclassing White Star's current fleet of liners. About a new fleet, Pirrie confirmed that changes in technologies would allow greater ships, up to 40% larger than anything currently available, to be built. Extra space on ships could be translated into greater luxury, a competitive differentiator over speed. So, effectively, this became the **strategy** to meet the **vision** of a fleet of superliners. Today, a *strategy* is a high-level view of how to attain the organization's vision in the future.

Figure 1.2. Lord Pirrie (chairman of Harland and Wolff) and Bruce Ismay (chairman of White Star) collaborated on creating a vision for the Olympic-class ships. [3]

Project Charter

Together, Bruce Ismay and Pirrie developed a **Project Charter**. The **business need** was to stem the loss of customers and drop in market share by improving the inferior service. The primary **business reasons for the project** were to win back customers and become the premier liner company again, which wholly supported the company's **strategic goals**.

17

The Project Purpose

This was to transform White Star's business model and replace the aging fleet with a new fleet of liners, the first new ships for a dozen years. The **project justification** was based on capturing the tourist class, which would be made of the nouveau riche, the North American industrialists spending their money taking a "Grand Tour" of Europe (Figure 1.3). Bruce Ismay and Pirrie reasoned this tourist class would make up the first- and second-class passages. Luxury, rather than speed of crossing, would drive these customers back. It was about the crossing's quality, the passenger or customer experience on board. The Olympic-class ships would be 40% larger than current ships with spaciousness, accommodation, and luxury. Speed would be sacrificed, and the ships would be slower than the competition's were. But it didn't matter about arriving on Tuesday night in New York if you arrived in greater style on Wednesday morning. The fleet would be built as Wednesday-morning ships where the passage was seven days rather than six days.

Figure 1.3. The British upper class and the American nouveau riche were the tourist classes Bruce Ismay targeted. [4]

White Star would run a 3-week schedule with two ships each departing Southampton every third Wednesday. Passengers would be picked up in Cherbourg, France, and emigrants in Queenstown, Ireland, arriving in New York on Wednesday morning. The ships were then provisioned and coaled during a 3.5-day stay. The ships would depart New York early Saturday morning, making landfall at Plymouth, crossing to Cherbourg, and then back to Southampton. Queenstown was used to load emigrants on only outbound legs.

The Business Case

This was based on a simple **cost benefit analysis** where passengers, cargo, and the delivery of mail (hence the name Royal Mail Ship) would boost revenues. During her maiden voyage, *Titanic* carried more than 3,500 bags with more than 500,000 pieces of mail. There would also be significant cost savings based on a reduced fleet size, from the current six ships, and running fewer (three) larger ships. Economies of scale would derive from a reduced workforce, less fuel, and lower maintenance costs.

The expected revenue could be readily calculated based on White Star's existing fare structure and that of the competition, Cunard Liners.

Passage	Expected Fare— 1911 British & US	Expected Fare— Today, US	Expected Passenger Numbers	Expected Revenue, US 1911, 100% Capacity	Expected Revenue, US Today, 100% Capacity[†]
First Class (Parlor Suite)	£870/$4,350	$83,200	Up to 100	$435,000	$8,320,000
First Class (Berth)	£30/$150	$2,975	805	$120,750	$2,394,875
Second Class	£12/$60	$1,200	664	$39,840	$796,800
Third Class	£3–£8/$15–$40	$298–$793	1,134	$28,350	$899,262
Total			2,603	$623,940	$12,410,937

Table 1.1. The Expected Fares of Passage by Class.

[†] *The expected revenue is based on a 100% fill capacity (unusual). The total maximum number on board, with a crew of 944, was 3,547*

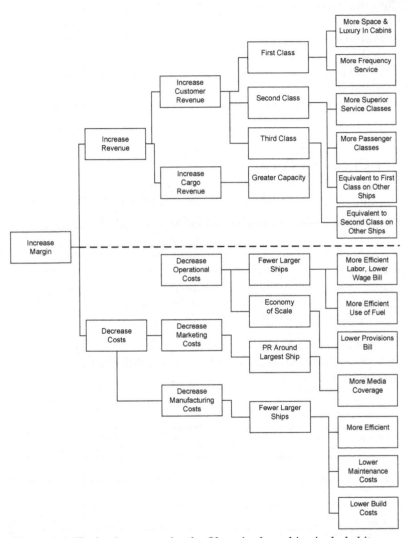

Figure 1.4. The business case for the Olympic-class ships included items for both increasing revenues and decreasing costs. For example, more advanced steam propulsion lowered coal consumption and required fewer stokers to feed the boilers, reducing labor costs.

The business case was very solid with a staggering 75% of total revenue based on first-class passage. As a result, the three liners would be paid for within a few years of going into operation. This was based on a break even point that would be reached in two years with a carrying capacity of 100%, or three years with 70% capacity (more realistic).

Project Funding

This would be provided by J. P. Morgan who is not immediately identified with the project. In 1907, J. P. Morgan had acted to restore order in the US financial system set off by a series of events known as the Panic of 1907 when the New York Stock Exchange fell close to 50% from its peak the previous year. J. P. Morgan pledged large sums of his money and convinced other New York bankers to do the same to shore up the banking system. J. P. Morgan had the financial clout and wherewithal to arrange the financing for the project through a major share issue.

Figure 1.5. The IMM Share Issue for White Star was used to fund the project for the Olympic-class ships. It raised £2.5 million for the project. [5]

21

Figure 1.6. 1902 editorial cartoon in Puck titled, "Following the Piper. His music enchants the world." It shows people of various countries and professions following the piper of big business, J. P. Morgan. It demonstrates the huge financial clout that J. P. Morgan carried and the stature of the man, the richest man in the world. [6]

With J. P. Morgan behind the project, confidence was sky high. The project was a major investment for White Star, as the liners were likely to be in service for at least twenty years. So, for the designers, it was critical to look to the future and get the design right, a ship that would not date too quickly. They proceeded with a design strategy of luxury over speed where the ship's second class equaled first class on other ships, and third class equaled second class.

The Project Goals and Objectives

The goals were to create three superliners to sweep the Atlantic, the ideal number, and to leverage the best in emerging technology currently available. The **project objectives**, which set the project direction and were a target for the project, were to deliver the Olympic-class ships over seven years. The delivery would be staggered so revenue from the first two ships could fund the third. The principal considerations for the ships were size, comfort, and luxury, with a reasonably high speed.

The Project's Criteria for Success Acceptance Criteria

These were that each ship had to be in operation within four years to pay back the large investments. They also had to be measurable. The **project's acceptance criteria** depended on meeting the conditions and terms of a contract, including each ship operating within specifications for weight distribution and loading details, speed under particular conditions of draft and deadweight, and seaworthiness and stability.

Project Constraints (External)

These had to be considered; for example, the weight or dimensional constraints were imposed, as the ships had to be built or maintained in a particular dry dock (see Figure 2.20). The limitations of piers also imposed dimensional constraints, as did the depth of dredged channels, which constrained the ship's navigational draft (vertical distance from the water surface to the lowest point on the ship). Constraints, for example, prearranged schedules and budgets, can restrict a project manager's options, but they can also include material availability, resource limits, and restrictions in the contract.

Organizational Constraints

These existed and had to be considered. For example, did Harland and Wolff have the ability to take the work on, a network of contractors, and the overall means to deliver the project? Environmental constraints existed but more on the working side once the project was completed. At this point, making and collecting assumptions started. These would have to be proved through various approaches and techniques, for example, the shipbuilder's model (see Figure 3.6).

Principal Stakeholders

Today, at this point in the project, we would **identify principal stakeholders** and sponsors (as part of **Human Resources Management**). The most significant stakeholder was sponsor J. P. Morgan, although he would stay in the background. Next was Bruce Ismay who would stay fully involved in the design and development. He would play a major role in the final project specification and in ensuring the maiden voyage was a success, which would help with continuing sales. Finally, Lord Pirrie would be the primary interface between Harland and Wolff and White Star.

Project Locations

The project was run across several locations. Harland and Wolff's chief shipyards were in Belfast, which included the design offices. White Star was headquartered in Liverpool with their main offices there. The new fleet would sail between a home port of Liverpool or Southampton, by way of Cherbourg and Queenstown, through to New York. The crew was sourced from the home port which had some repair facilities.

For more information about Bruce Ismay and Lord Pirrie, see Appendix A.

Chapter Wrap-up

Conclusion

The project was begun on a very sound footing. White Star faced a pressing business need to replace their aging fleet of liners to have any chance of competing with stiff growing competition. White Star embarked on a well-thought-out strategy to invest in new emerging technologies and create three superliners. White Star's relationship with Harland and Wolff was very important at this point in the project's initiation. White Star saw no need to tender this critical contract.

Key Lessons

Today, before a project is started, stakeholders and the potential project team would:

- Make a "go/no-go" decision on whether your solution will be workable, that is, it has enough value to pay for and support itself and is not a substantial "risk" to the business.

- Go through a quick cost-benefit analysis to highlight its viability and draw on a payback period. Ensure the project team understands it. At this point, few projects go through a more detailed business case to forecast a return on investment and calculate the risks during the project and once the project output is in production.

- Ensure the project charter, stakeholders, and sponsors are in place. Identify any conflicts of interest early in the project.

- Establish a budget.

Other aspects to consider are identifying:

- Project's goals, objectives, purpose, criteria for success, critical success factors (these might vary by stakeholder group).

- Any constraints and assumptions.

- All principal stakeholders and sponsors, using a stakeholder map. Stakeholder expectations should be identified, documented, and prioritized (known as stakeholder analysis).

    ○ The output is a stakeholder register, a directory of all the stakeholders and their stance on the project (supportive, negative, or neutral).

- The principal customers or target audience and their wants, needs, and expectations.

- A feasibility Study.

Educators

Discussion points:

- White Star's business case was very solid with a short return period. But was this enough, and did it cover the necessary risks adequately, did it look at the bigger picture, the economic environment in the longer term?

- Discuss White Star's relationship with Harland and Wolff and the pros and cons of having Lord Pirrie on the White Star board of directors.

- Can a good buyer/seller relationship accelerate a project's start-up time?

- Should White Star have gone to tender with the initiative for second opinions?

CHAPTER 2

Planning Phase

In This Time Frame

○ March 1908—Announcement of the project to the press.

Overview

This chapter looks at how White Star and Harland and Wolff planned the project. Today, at this point in the project, the project management plan's development integrates all the subsidiary plans from the various knowledge areas into one cohesive whole. These include plans for Scope, **acceptance criteria**, and **constraints**.

Integration can occur outside and in the project, an important consideration for Harland and Wolff, as is Requirements Management, Time (Schedule) Management, Cost Management, Quality Management, Process Improvement, Human Resources Management, Communications Management, Risk Management, and Procurement Management.

Integration Management

White Star proceeded with the development of the project management plan, which typically covers many activities (from chapter 1) such as the **project charter**, **business case**, **objectives**, **criteria for success** had other ships under

27

construction, and this could influence the project, as typically facilities, equipment, materials, and workforce were shared to optimize the projects through economies of scale.

In today's projects, **Integration Management** processes are prevalent and include **direct and manage project execution, monitor and control project work, perform integrated change control,** and **close the project or phase.**

Scope Management

The planning phase typically starts by scoping the project. When it came to **Scope Management,** the project collected requirements (business or shipowner) from White Star and the principal stakeholder Bruce Ismay. These were to do the following:

- Deliver three identical superliners with priority on spacious accommodations and luxury (especially in first class).

- Improve existing service for all three classes where second class equaled first class on other lines, and the third class equal to second class on other lines.

Requirements Gathering

This was part of a requirements-gathering process that determined and defined the features and functions. For Harland and Wolff, these had to meet the needs and expectations of the sponsor, client, and other stakeholders (for example, government regulators). Harland and Wolff had an eighty-year business relationship with White Star. The chairman Lord Pirrie was on the White Star board, and he would help interpret the White Star requirements for his design team. Requirements that were more detailed would be collected in the design phase. Pirrie and his design team would go through a **progressive elaboration**, the progressive improvement of the plan as details that were more particular became available during the project. All projects have boundaries defining what is included in the project scope and what is excluded (the **scope statement**).

In Scope

In defining what was **in scope** for the project, the approach was very much based on experience with previous ships and the tried and tested approach of keel-and-ribs construction but significantly scaled up by the project. Harland and Wolff reached into its **organizational process assets** for these formal processes and best practices in shipbuilding. Harland and Wolff also harnessed their **expert judgment** based on their expertise with this type of hull design known as the Belfast Bottom. This flat-bottomed hull allowed an increase in the ship's length without the need to enlarge the width.[1]

Out of Scope

It was also important to determine what was **out of scope** for the project, for example, an extremely fast ship that could compete on speed with the competition. This one factor affected very much the power requirements and the propulsion system. The identification of out-of-scope items is critical, as is the communication of these to the project team, as this can be very costly if not completed.

Technologies Impact on Scope

The scope was very much affected by the introduction of emerging technologies such as automated control systems, advanced steam propulsion systems (high-pressure steam reused several times), wireless communication systems, and electrification of the entire ship. These technologies added complexity to the project, and some required a learning curve for the project team. They were also continuously changing. For example, most of these technologies were proved and had been used in other ships, but this was the first time they were all put together or integrated into one package.

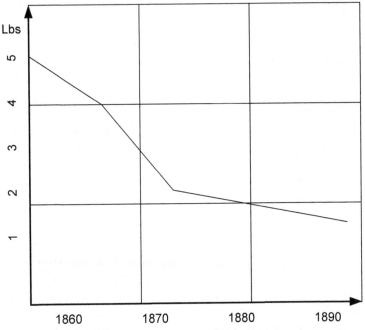

Figure 2.1. Coal consumption of steam engines per horsepower per hour shows improvement in efficiency, which improved the "distance margin" and speed of steamships. Improvements in marine steam engines included the boilers, through better quality steel, and better lubricants that improved airtight sealing, hence facilitating higher steam pressures.[2]

Third-Class Passengers

With the attention on first class, it was easy to forget that there was an opportunity to improve third-class passage radically, so it would attract booming immigrant traffic. Immigration into North America from Europe was up to one million immigrants per year. Typically, taken up by the working classes, third class was not expected to have any luxury because many of these passengers did not have a running-water toilet in their homes. Bruce Ismay brought the third-class (steerage class) passage **into the project scope** but was cautious of its influencing overall space, so minimal space was allocated to third class (7%), compared with the first-class allocation (60%), which did not deter the typical third-class passenger.

Figure 2.2. Immigrants arriving at Ellis Island. One million a year or up to five thousand a day. This was a lucrative trade for the liner companies such as White Star.[3]

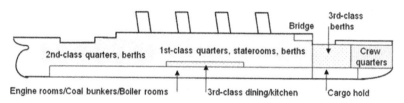

Figure 2.3. Envisioned space allocation outlines third class (7%) compared with first class (60%).

Product Analysis

A project of this magnitude required a vast quantity of materials and products at various points in the project. At this point, a **product analysis** identified the full bill of materials required and considerations for materials buyers needed to purchase. The **lead time** on ordering some of these was extensive and had to be scheduled (see **Time Management**).

The Project Scope Statement

One of the most important outputs of scope management is the **project scope statement,** as it is an important project reference document, the basis for any future project decisions, and it is revised to reflect approved changes to project requirements. It agreed the project scope between Harland and Wolff and White Star with details of the characteristics of the ships, deliverables, exclusions, constraints and assumptions, and acceptance criteria.

Project Work Breakdown Structure (WBS)

This is a subdivision of effort where **decomposition** is used to create a set of deliverables. Further into the project, this is further decomposed into smaller **work packages** (at the lowest level) which are more accurately estimated, more easily assigned to individuals, and tracked.

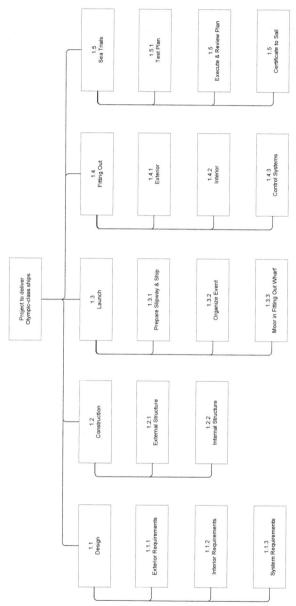

Figure 2.4. Projection of what a possible project WBS would have looked like for the project. The WBS is one of the most important project management tools available today that helps build team consensus and buy-in to the project. It is also a control mechanism to keep the project on track.

In today's projects, a **scope baseline** provides the approved detailed project scope statement, with the WBS and a WBS dictionary that defines the WBS parts.

Time Management

When it came to **time management,** the project followed established lines and a traditional approach. The WBS was decomposed into a set of activities. The **sequence activities** process organized the project activities in the right order. Most of these were tested over time (best practices). For example, in the construction phase, the approach consisted of laying a keel and attaching a series of ribs. To these, large (6 feet wide and 30 feet long, weighing 3 tons) steel plates were riveted, which was the most significant project activity.

Estimate Activity Resources

With the sequence of project activities organized in the right order, estimating activity resources referred to not just the workforce, but also equipment, facilities, and materials. The project teams building the various ships shared the equipment and facilities. Creating an accurate project schedule requires accuracy in determining the resources (skills, quantity, and availability) to complete the activities. These estimates are then used to estimate project costs. Through experience, Harland and Wolff had a very well-defined mapping of resources to activities, notably, the demarcation of work through the trades (see **human resources management**), and which skills, within a trade, would be needed during construction or fitting out.

Estimate Activity Duration

This was the next step, which defined the time in the number of work periods to complete each activity. Harland and Wolff had to work with new emerging technologies, but most were proved and had been used in previous ships. Therefore, the estimates were **deterministic** (a single estimate was used where the duration was known with a fair degree of certainty). Estimates on the integration into one package were **probabilistic** in that the duration was uncertain. In today's project, a **three-point estimating** technique or weighted average would be used.

Cost Management

This is a three-step process composed of estimating costs, determining a budget, and controlling costs. When it came to **cost management**, the project faced a solid business case and return. As a result, the project had the leeway to push for the best in creating the world's most luxurious ship, if it was delivered on time.

Estimating Costs

Harland and Wolff shipyards had to **estimate costs** accurately. The project work had been done before in previous projects, so the confidence level in the duration estimate was high. The main technique of the time was for shipyards to base **analogous estimates** on their experience in building similar ships of nearly the same type, but not size. The Olympic-class was a scale up in size from White Star's previous Big Four: *Celtic, Cedric, Baltic,* and *Adriatic.* With previous estimates, the expert knowledge and self-made formulas (a form of **parametric estimating**) would have been used but kept secret and not put in the public domain. Therefore, estimating labor costs was analogous, as was materials cost estimation. Both of these were done repeatedly, with all the shipbuilding continuing at the shipyards, so estimating was straightforward for the estimators.

As the workforce operated in trades, much of the work they did was repetitive from ship to ship. Building three identical ships had an advantage that the costs decreased with time as workers became more efficient, as part of their **learning curve**, and this was factored into the estimates.

Today, to mitigate risk over the estimates, a project manager can use a **three-point estimate** where cost estimates are created for each activity, namely, optimistic, most likely, and pessimistic. An average is created for the overall cost.

The most expensive costs were related to the new emerging technologies with the advanced steam propulsion systems (turbines), automated control systems, and communications or Marconi Wireless Telegraphy equipment (system). Although new, most of these technologies were already proved in other White Star ships. Another significant cost was fitting out the first- and second-class quarters, which required craftsmanship (available through the trades) and materials of the highest quality.

For the deliverables, Harland and Wolff completed a final estimate of the costs, determined the profitability margins from which they could announce a final fixed price of £3 million ($15 million) for the pair of ships. This was agreed at the time of contract signing with White Star, see **procurement management**. It was just under the project **fund limit**.

Determining a Budget

With the experience Harland and Wolff had, unforeseen costs were unlikely. To create a budget, the process involved aggregating the expected cost the project would incur to create an authorized **cost performance baseline**, which was then used throughout the project to measure project performance. This was the authorized Budget at Completion (BAC) spread over the project schedule. Typically, both **contingency and management reserves** were established, although not part of the cost performance baseline, to account for uncertainty and risks and for any unplanned changes. Therefore, with reserves in place, staying on budget would be straightforward for Harland and Wolff.

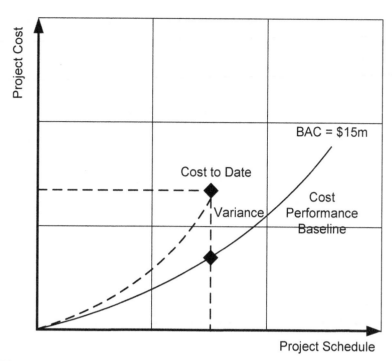

Figure 2.5. Costs were assigned to the work in the project to create an authorized cost performance baseline for measuring project performance. This involved calculating the Budget at Completion (BAC).

Controlling Costs

This was performed regularly by monitoring the planned value, actual cost, and earned value. The Harland and Wolff board of directors, at their directors' meetings, were given a status of the project for the pace of work and the actual costs, which equals a **performance report** to indicate how well the project is going. Today, costs are tracked by an S-curve, a graphic representation of the accumulated budgeted costs over time, through the project life. Figure 2.6 below represents this, outlining the costs (planned versus actual) early (the design phase) in the Olympic-class project. It also outlines the Budget at Completion (BAC), £1.5 million ($7.5 million) per ship; the estimated cost of the project when completed; the project baseline; the Estimate at Completion (EAC); and the amount expected for the total project to cost, on completion.

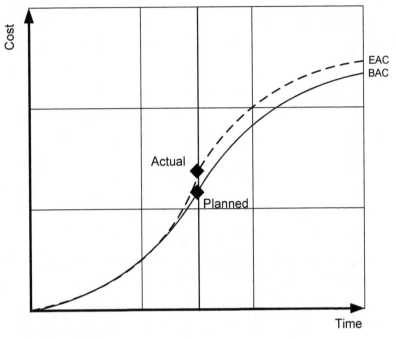

Figure 2.6. An S-curve of the early part of the Olympic-class project and the projected costs (actual vs. planned). The vertical bar represents a point in time. Typically, these costs were tracked monthly or even weekly, if the project was not going well.

If **performance reports indicated** the project was going poorly, operating beyond its budget, or off schedule, this would have triggered change requests to increase the workforce, add corrective actions, or start quality activities.

Quality Management

When it came to **Quality Management,** the project strove to create a world-beating customer experience for crossing quality. The **market expectations** of the project were inevitably going to be very high, with the way the ships were marketed, and this helped shape the overall approach to quality. It also helped determine what preventive measures had to be taken to avoid nonconformance costs (rework or scrap). The shipowners' requirements defined the following:

- Salability—a balance of quality and cost in the fares for all three classes.

- Constructability—the ability for Harland and Wolff to build the ships with the available technology and workers at an acceptable cost.

- Social Acceptability—the degree of conflict between the ships and the values of society. These ships were a source of pride of society for Belfast, and the United Kingdom (100,000 people would witness the launch). They were technological achievements of their time.

- Operability—how much the ships could be safely operated.

- Availability—the likelihood that the ships, under given conditions, would perform satisfactorily. The key parts were:

  ○ Reliability—the likelihood that the ships would perform, without failure, under given conditions for a set time.

  ○ Maintainability—the ability to restore the ships to their stated performance level within a specified period.

Quality Management Plan

This typically identifies the **quality standards** and determines how these would be satisfied, that is, how the project will be measured for compliance to these. One first step is to look across the project (today, the WBS is used) to identify the specifics that will satisfy the project requirements for deliverables and where quality would have the most impact. Areas of particular importance would have been the ship interiors, especially those very visible to the passengers.

As the project mantra was for customer experience and the quality of crossing (satisfaction and winning repeat business), the **quality standards** had to be very high and exceed the competition's offering (Cunard). This was definitely the case for the first and second-class interiors and stretched to the third class, for example, the cabin or stateroom accommodations, the public areas, the large staterooms, and new features such as the swimming pool. Harland and Wolff used White Star to confirm formally whether these standards were met. This would be addressed in the fitting out phase.

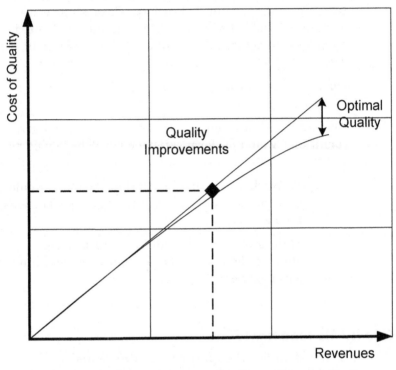

Figure 2.7. There is always a cost to quality. Preventing mistakes is always more cost effective than correcting them. Marginal analysis compares the cost of incremental improvements to the deliverable (ocean liner) against the likely increases in revenue.

Quality Standards

Harland and Wolff reviewed the shipping industry's standards and regulations to determine that both the project plan and the plan for quality were acceptable. For example, the project had specific regulations it had to meet. The regulations' relevance must be planned into the project to conform to requirements.

An example of quality standards was embedded in the **procurement process** for rivets, steel plating, and all the steel on boilers, bulkheads, and watertight doors. Harland and Wolff left it to the suppliers to prove their deliverables met the quality standards (see Figure 2.8). These materials were also inspected on arrival at site.

The cost of iron shipbuilding had decreased rapidly from the mid-1870s because of better metalworking machine tools and cheaper metals. Although Harland and Wolff followed this trend, they had to balance this against the quality standards set for the project.

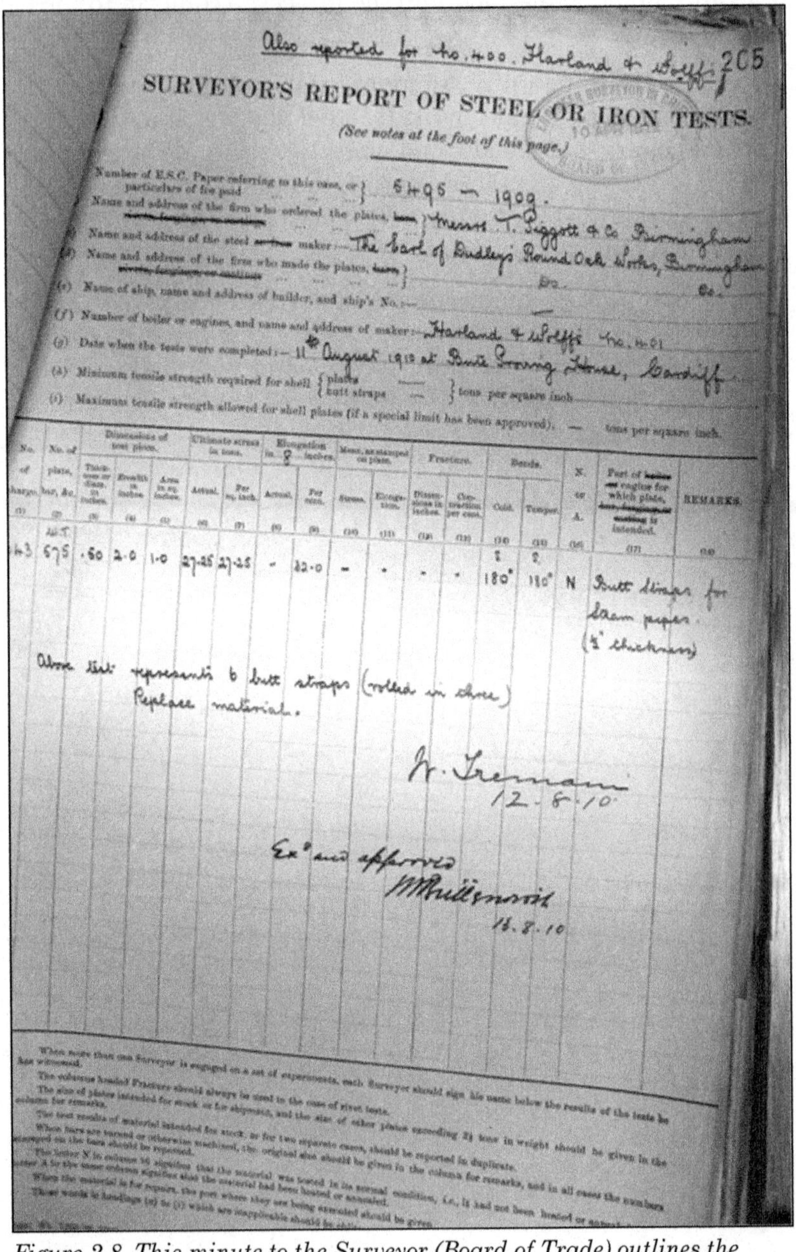

Also reported for No. 400. Harland & Wolff 205

SURVEYOR'S REPORT OF STEEL OR IRON TESTS.

(See notes at the foot of this page.)

Number of E.S.C. Paper referring to this case, or particulars of fee paid: 5495 – 1909.

Name and address of the firm who ordered the plates: Messrs T. Piggott & Co Birmingham

Name and address of the steel maker: The Earl of Dudley's Round Oak Works, Birmingham

Name and address of the firm who made the plates: Do. Co.

Name of ship, name and address of builder, and ship's No.:

Number of boiler or engines, and name and address of maker: Harland & Wolff No. 401

Date when the tests were completed: 11th August 1910 at Bute Proving House, Cardiff.

Minimum tensile strength required for shell plates / butt straps — tons per square inch

Maximum tensile strength allowed for shell plates (if a special limit has been approved). — tons per square inch.

No. of charge	No. of plate, bar, &c.	Thickness	Breadth	Area	Ultimate stress Actual	Ultimate stress Per sq. inch	Elongation Actual	Elongation Per cent.	Stress	Elongation	Dimensions in inches	Contraction per cent.	Bends Cold	Bends Temper	N. or A.	Part of engine for which plate is intended	REMARKS
43	W.T. 675	.60	2.0	1.0	27.25	27.25	-	32.0	-	-	-	-	180°	180°	N	Butt straps for Steam pipes (⅝" thickness)	

Above test represents 6 butt straps (rolled in three). Replace material.

W. Tremain 12.8.10

Exd and approved
W. Bullimore 15.8.10

Figure 2.8. This minute to the Surveyor (Board of Trade) outlines the testing done by the supplier (in Cardiff) that had to comply with the quality standards set for the project.[4]

The workforce was perceived as, and they saw themselves as, craftsmen operating within a system of trades. For these workers, reaching set quality standards was very important for their individual reputation and to that of the trade to which they belonged.

Quality Assurance

These are regular structured reviews to ensure the project follows the planned quality standards by using various collected measurements. The Board of Trade inspectors who made regular **external inspections** during the project's execution completed their activities.

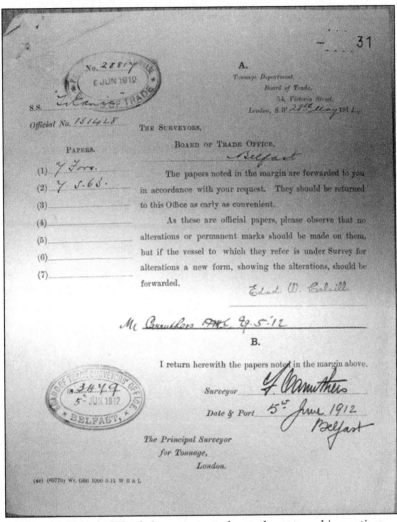

Figure 2.9. Board of Trade inspectors made regular external inspections during the execution of the project to collect various measurements and determine whether quality standards were met. Harland and Wolff was a premium shipbuilder in the world that had a reputation for quality. This was a strong factor in their selection as shipbuilder to White Star.[5]

In the project, the determination of quality relied not only on **inspection methods**, but also on **prevention** to keep problems from ever occurring, for example, by using high-quality parts and materials. Harland and Wolff strove to get the best of these to secure a lasting reputation. However, these were not always readily available, as there were shortages, for example, of top-quality iron available. Harland and Wolff dealt with the shortages by using a grade down from the best-quality iron, known as No. 4 or "best-best." The company used No. 3, graded "best." This is perfectly acceptable as **quality** and **grade** are different, where quality is the sum of the characteristics of the ship that meet project expectations. Grade is a category for entities (such as rivets) with the same functional use, but different technical characteristics. Lower grades are acceptable as long as the grade of an entity is assessed for its potential impact on quality.

Quality Control

This was done through inspections and was done within the trades as the work was completed. A good quality control system selects what to control, sets standards, establishes measurement methods, compares actuals with standards, and acts when standards are not met. The trades would work within the shipyard to set the standards and measurements. They would also complete the inspections. For example, one of the most challenging inspections was for the Rivet Counter who had to inspect completed rivets to ensure tight fit. A loose-fitting rivet could lead to leaks. The inspections were very challenging because of the volume—more than 3 million rivets.

In today's projects, we would use **quality control tools** such as cause-and-effect diagrams, control charts, flowcharts, histograms, Pareto charts, run charts, scatter diagrams, and statistical sampling.

Iron Triangle

The cornerstone of any project is the **Iron Triangle** of scope, time, and cost. With the project, the most fixed constraint was time. The maiden voyage was a hard deadline. The scope was fixed as well. Quality could not be compromised because of White Star's reputation. Therefore, the only flexibility lay in the cost and resources.

Figure 2.10. The **Iron Triangle** *is the cornerstone of project management and is a careful balance of cost, scope, and time, with quality at the center.*

In today's projects, one approach is to use **control accounts,** which are **management control points** where the integration of scope, schedule, and cost occurs and where performance is measured. A control account typically controls several work packages (each valued at say 500 hours), where the sum of the control accounts adds to the total project value.

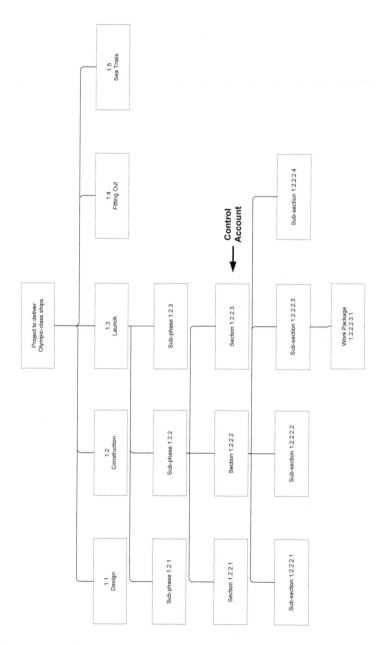

Figure 2.11. The control account typically controls several work packages (each valued at say 500 hours) where the sum of the control accounts adds to the total project value.

Human Resources Management

When it came to **Human Resources Management,** a well-rounded team was required for the project and the maiden voyage's implementation. Today, the human resources plan provides guidance to the project roles and responsibilities, organization, staffing, and how people will be managed, brought in, and released. For this, the plan uses a **responsibility assignment matrix**, a staffing management plan (a resource histogram; see Figure 5.14), and a project organizational chart (**organizational breakdown structure**).

Project Staffing

Harland and Wolff provided the design team (naval architects, designers, and draftsmen) and a large construction workforce. Harland and Wolff could expand the workforce and get and develop workers as needed by the project demands (more next chapter). White Star staffed the project with the operations team who had to accept the deliverable and take it through sea trials. This team was composed of the captain, the senior officers, and engineers. For the maiden voyage, a crew capable of running the ship under the officers' auspices had to be added to the team.

For the project, Bruce Ismay took on a marketing role, principally to ensure the maiden voyage's success. He also wanted to see that the customer experience had priority, and it emerged as part of the project mantra for the rest of the project.

Communications Management

When it came to **Communications Management,** the first step was to identify all the project stakeholders, who ranged from the project team to the end customers, the classes of passengers, and to the media. Bruce Ismay believed he understood the aspirations of his customers, particularly the first-class passengers, as he moved in those circles. The first class had tremendous power, influence, and sway in the new fleet's success, so an appropriate communication strategy had to be developed for them.

Communications Management Plan

This documents the information needs of stakeholders in who needs what, when, how they will get it, and from whom. A project uses several communication channels such as up to management, down to subordinates, and lateral to team members, functional groups, and customers.

Internal and External Communication

This dealt with principally the project team and the Harland and Wolff board, to whom the project manager would have reported. The focus was project status. **External communication** dealt with the media and the public, and it started at the project's outset. The focus was on promoting the project and outlining its scope for the ships' size and the expected level of luxury. It also delivered a new fleet for White Star, thus putting them ahead of the competition. Hence, everyone in the project understood the project's goals. This also gave a common reference point to all stakeholders, important later when the project manager had to manage these stakeholders. The name Olympic-class was chosen to inspire a sense of awe as shown by the announcement to the press, and in the *New York Times* of March 1908 (see Figure 2.13 below). Unlike the competition, White Star did not keep the names of their new liners a secret until the launch.

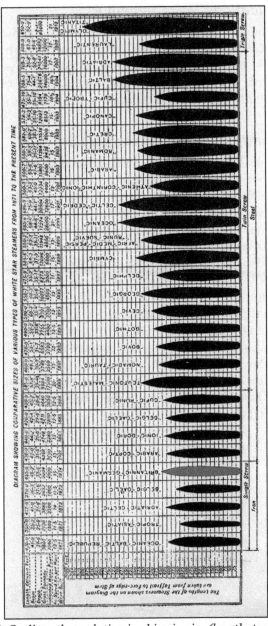

Figure 2.12. Outlines the evolution in ships in size (length, tonnage, propulsion, materials) and the comparative size of types of "White Star Steamers" from 1871 to Present. The First Industrial Revolution set the public expectation to the progressive evolution of technology in size. [6]

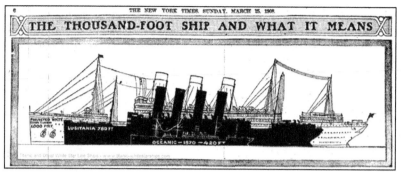

Figure 2.13. The New York Times announcement highlighted the scale up in ships from the current line to the competition to the planned. In twelve years, ships had scaled up by 400% in size. It also underlined the project's scope, and it was designed to inspire the audience.[7]

Marketing Role for the Project

When Bruce Ismay took on a marketing role for the project, he did so because, as part of the upper class, he moved in the right circles of the British aristocracy and the American nouveau riche. He would be able to communicate effectively with this target audience. He could promote or sell the maiden voyage and ensure it was filled with important people, which would set the precedent for future voyages, as friends would sell the experience of their voyage to other friends.

Risk Management

When it came to **Risk Management,** both White Star and Harland and Wolff had vast experience in building and operating ships across the transatlantic route. With their long history and experience, both companies knew the risks in such a large project. It is safe to say that both were conservative, risk-averse companies. However, the strong project financial returns would have encouraged their overall enthusiasm for the project and increased the acceptable risk level. Both companies were responsible for risks in their domain of responsibility, that is, in the project and in the operation.

In today's projects, one of the first activities is to determine the threshold of risk a project should handle (see Figure 2.14), that is, how much authority is given to a project manager before there is an escalation. The escalation should be through a control board, to determine costs, with a path to sponsors and senior management.

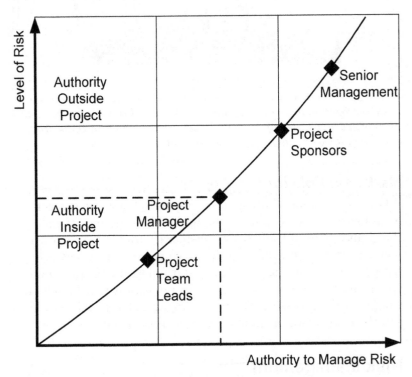

Figure 2.14. Outlines the boundaries of risk and the items authorized to a project. The level of risk a project is authorized to deal with should be predetermined.

The project risk threshold is very much based on the sponsor's (and possibly stakeholder's) risk tolerance. This is driven primarily by the priority of the project within the company, which, for White Star, was the most important company-wide. There is also a cost to decreasing the risk, driven by the project's financial return (see Figure 2.15).

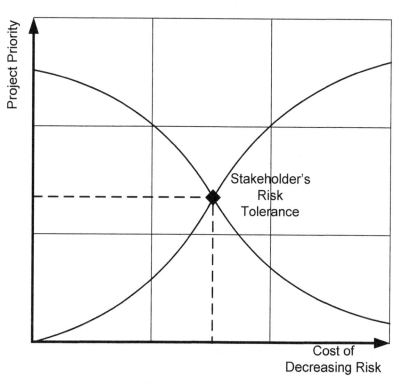

Figure 2.15. The cost and effort to decrease risk (chance of failure) is in proportion to the stakeholder's tolerance of risk (and sponsors) on the project, which is affected by the project's priority. Bruce Ismay's confidence in both White Star and Harland and Wolff increased his risk tolerance, as did the strong financial returns from the project.

YEAR	SHIP	LINE	DEATHS
1854	*Arctic*	Collins	278
1856	*Pacific*	Collins	186
1857	*Tempest*	Anchor	150
1868	*Hibernia*	Anchor	66
1870	*Cambria*	Anchor	190
1873	*Ismalia*	Anchor	52
1873	*Atlantic*	White Star	546
1895	*Elbe*	North German Lloyd	303
1998	*La Bourgogne*	French	549
1904	*Norge*	United Steamship Co.	701
1907	*Berlin*	North German Lloyd	140
1909	*Republic (& Florida)*	White Star	4
			3165

Figure 2.16. White Star knew well the risks in the North Atlantic. A catalogue of ships lost in disasters, for example, going onto rocks or full speed in bad weather, in previous years provided a rich history from which to work.

Categories of Risk

In today's projects, we would use categories including **external risks** such as regulations, customers, and market conditions; project management risks such as poor estimates or a novice project team; organizational risks such as delay in staff or fund availability or staff conflicts; and technical risks such as new technology or quality concerns.

Identification of Risks in the Project

White Star might not have used the same categories, but the most significant were related to getting the financials right (the costs versus the return on investment) and selecting the right integrator to meet contractual obligations. The latter would have to be under the stipulation of a formal contract.

In this situation, the contract was used for insurance purposes, the demarcation of responsibilities, and the final acceptance (of deliverables). Hence, White Star was transferring some risk through contracts for the responsibility of project risks.

For Harland and Wolff, the **identification of risks** was related to those principally in the project. These included defining an accurate scope (functional and nonfunctional requirements) and schedule to meet business priorities and delivery within the predefined budget. Other risks were technical, related to technology, like new emerging technologies or the integration of various control systems (docking, steering, maneuvering, communications, fog warning) to a single point. In the project, there was also the risk of not meeting government regulations. A vendor such as Harland and Wolff would have had the means to be fully aware of these risks to manage these by bringing the right project team to the table.

In today's projects, one approach is to identify risk categories by using a risk breakdown structure (RBS) (see Figure 2.17).

Technical (Quality, or performance risks)	External (Outside of the project)	Organizational (Unreasonable cost, time, and scope)	Project Management (Faults in managing project)
Use of unproven and complex emerging technology	Select wrong integrator to meet contractual obligations	Inadequate funding or the disruption of funding	Inaccurate scope (requirements functional/non)
Ability to integrate technologies to a single point	Legal issues-regulations	Unreasonable cost, time, and scope expectations	Inaccurate financials (costs/return)
Expectations for impractical levels of quality and performance	Labor issues (Unions)	Poor project prioritization	Inaccurate schedule and resource requirements
Changes to industry standard	A shift in project priorities for White Star	Competition with other projects (ships) for internal resources	Low-quality work or not addressing defects adequately
Not meeting government regulations	Weather "Force majeure," risk in the Atlantic, storms, traversing "Iceberg Alley."	Operational readiness and preparedness of officers and crew	Not transferring lessons learned between ships
	Certain months like April worst months for icebergs		Failure to test and inadequate sea trials
	Long construction project (6 yrs) Changes in business model, technology, or events		
	Poor External Communication		

Figure 2.17 A sample risk breakdown structure categorizes the project risks over four categories.

Identification of Risks in the Operation

White Star had a comprehensive understanding of the risks in operation based on their vast transatlantic experience. These risks included the risks in the Atlantic, storms, and traversing "Iceberg Alley." Some months, such as April, were the worst months for icebergs. In a project of this magnitude, identifying all risks in operation was important.

When it came to these risks, the most significant for White Star were the operational readiness for the maiden voyage and the preparedness of officers and crew. A worst-case scenario would relate to the loss of service, or even a disaster at sea. White Star, like any shipping company today, took out marine insurance that covered most possibilities in a disaster. But what could the loss of a superliner mean to White Star? Could it endanger the company? A worst-case scenario like a disaster, where White Star was found negligent for loss of life, could put the company out of business through the resulting lawsuits alone. In creating a comprehensive business case for your project today, you need to factor in this risk.

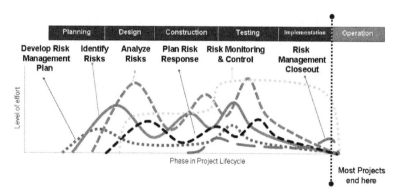

Figure 2.18. In today's projects, risk management should be continuous through each project phase. For example, identifying risks does not stop in the planning phase, as does the risk response plan, which evolves.

Risk Analysis

This was affected by the project status where early, many risks had not surfaced, for example, those related to some emerging technologies and their integration. As the project proceeded, new risks became more evident and went through analysis (see Figure 2.18).

Risk analysis requires information and recurring (repeating) projects have good historical information to work with, in contrast to first-time projects. This project was unique, had never been delivered, in scale of ships and the installation and integration of some emerging technologies, although there were some aspects repeated and familiar to Harland and Wolff.

The relationship between White Star and Harland and Wolff was very tight, developed over time, and in many ways, this helped reduce the risk. For example, Harland and Wolff had great expertise in technologies and could quickly assess options. Both parties had experience in working with each, and they could better anticipate their individual strengths and weaknesses.

Today, projects go through a **qualitative risk analysis** where risks are scored and ranked based on their probability and impact. This is followed by a **quantitative risk analysis** that assesses numerically the probability and impact of the identified risks and creates a risk score for the project. More in-depth than qualitative risk analysis, it provides a prioritized list of risks and the trends in the identified risks.

Planning Risk Responses

For both White Star and Harland and Wolff, a contract was a very significant mechanism for responding to risk, particularly for White Star, to ensure the integrator managed project risks. In this situation, the contract was used for insurance purposes, the demarcation of responsibilities, and the final acceptance (of deliverables). Hence, White Star transferred some risk through a contract for the responsibility of project risks.

One optional strategy was falling back on the project sponsor J. P. Morgan and the significant financial reserves available to him. However, Morgan had been patient, and since his takeover in 1902, he now looked for a return on his investment.

Today, the approach to **planning risk responses** is to decrease the possibility of risks affecting the project adversely and on maximizing positive risks to help the project by assigning responsibilities to project team members close to the risk event. The response should be appropriate to the risk in cost and time invested. The project team individuals assigned to the risk must have the right level of authority.

Procurement Management

When it came to **Procurement Management,** the project required extensive procurements up front and then throughout the extensive schedule of the project to construct the two Olympic-class ships. White Star had a long-term agreement with Harland and Wolff not to contract with any rival shipbuilders. Harland and Wolff were the preferred shipbuilders for White Star, and they would never build a ship for any of White Star's competitors. The agreement had worked well for both parties over the years.

Planning Procurements

Harland and Wolff were primarily responsible for planning the procurements and had formal purchasing procedures that had to be followed. The process for **Planning Procurements** relied on the project scope, requirements, risks, schedule, availability of resources, quality, and cost. A **make-or-buy analysis** determined what would be bought externally or produced in the Belfast shipyard, which was well equipped with production capacity and a skilled workforce.

Project Contract

This was for Harland and Wolff to deliver two Olympic-class liners. In this type of contract, it is typical to distribute the risk between both parties so they have the motivation and incentives for meeting the contract goal. Only Pirrie at Harland and Wolff knew the precise contract conditions and its financial outcome. The contract was delivery oriented with a cost-plus basis where the bill of materials and labor were passed on to White Star. However high the project costs became through specification, changes, or increases in material costs, Harland and Wolff were guaranteed a 5% profit on contract, which was an incentive. There were penalties and fees on the deliverable, but the contract was very much a standard practice in the shipbuilding industry. The downside of cost-plus contracts is that they did not provide any incentive for the shipyard management to save costs.

Investing in the Project

In the bigger picture, the project was a great opportunity for Harland and Wolff to reinvest and re-configure the shipyards' infrastructure, which included getting gantries and cranes. Harland and Wolff turned to the engineers who built the Firth of Forth Bridge in Scotland for the new gantries that could support a complicated crane system (central and revolving) to reach every part of the ship. The investment costs of £100,000 were substantial, as these gantries were the largest built in length, height (840 feet long by 240 feet wide), and capability (four large electric lifts). Harland and Wolff also procured a 200-ton floating crane from Germany for £30,000 to lower the ships' massive engines once the ship had been launched and sitting in the water, which was a very complex process.

The size of the ships imposed dimensional constraints on the existing facilities, so new ones were required. The first, a large outfitting wharf (Thompson), was constructed, and water was dredged to a depth of 32 feet. Here, during the outfitting phase, the ships' engines, boilers, and superstructure would be added to the hollow shell, and work was completed on the interior to turn it into a liner.

The second, a new dry dock, was required for maintenance or emergency repairs. Fortunately, a dry dock, the Graving Dock, was available. In 1904, with much foresight to how ships were increasing in size, work had begun on the 900-foot long dock, whose walls were 18.5 feet thick and had 332 massive keel-blocks of cast iron to support the weight of the great liners it would hold. This dock had to be extended so its first ship, *Olympic,* could enter in April 1911. One of the principal features was the speed with which the dock could be filled and emptied in under an hour, important in ensuring a quick turnaround to keep with the schedule.

This entire new infrastructure modernized the shipyard and equipped Harland and Wolff with a capability to build much larger ships. It put them in a good position for future business and further orders.

Figure 2.19. Harland and Wolff heavily invested in the shipyard by constructing two massive slip-ways and gantries into the Lagan River. The slip-ways slope away from the new gantries that are being erected.[8]

Figure 2.20. The massive new Graving Dock (dry dock) at the Belfast shipyards equipped Harland and Wolff with the capability to handle much larger ships. This complex included a sophisticated pump house that could empty or fill the dock in less than an hour.

Figure 2.21. The announcement proudly outlines the scale of the expansion and reorganization of the shipyard in 1907 with extensive workshops and engineering facilities.[9]

Chapter Wrap-up

Conclusion

The project was planned with great due diligence. The scope was not seen as daunting, but very achievable, based on the previous project experiences of White Star and Harland and Wolff. Significant attention was paid to all the PMBoK® Knowledge Areas we use today, particularly the iron triangle of scope, cost, and time. The project very much depended on the strong business relationship between White Star and Harland and Wolff. Both parties, particularly Harland and Wolff, had to invest in massive upgrades to their shipyard facilities.

Key Lessons for Today

Today, typically before project managers commit to a project, they would do the following:

- Ensure due diligence in defining the business problem and competitive services

- Determine by segments customer/target audience and value propositions and create profiles and scenarios for these

- Define the scope, cost, and time of the project and pay much attention to the iron triangle of project management (Figure 2.10).

- Pay attention to all nine PMBoK® Knowledge Areas, and in particular,

  ○ Ensure a control mechanism is established early to control the scope, as it can continue to expand

  ○ Ensure the human resources plan outlines training plans, not just for the project team, but also the operations team

  ○ Identify how critical success factors are viewed by various stakeholder groups

Educators

Discussion points:

- Discuss the project phase through the lens of each of the nine PMBoK® Knowledge Areas of Scope, Time, Cost, Quality, Process Improvement, Human Resources, Communications, Risk, and Procurement.

- Discuss the relationship between White Star and Harland and Wolff in light of the above Knowledge Areas and the planning process. Were risks adequately identified?

- How well was the project phase completed?

CHAPTER 3

Design Phase

In This Time Frame

○ April 1907—Order signed to proceed with the design of the two new liners.

○ June 1907—Design work started at the Harland and Wolff shipyard.

○ July 1907—Work started on two new slip-ways under the overhead Arrol gantry.

○ June 1908—Board of Trade approval on the project's general specifications.

○ July 1908—Final designs approved by White Star directors.

Overview

This chapter looks at how White Star proceeded with the design phase to create a detailed design and update the various plans (for the knowledge area), notably scope and time management. Today, at this point in the project, requirements that are more detailed are defined from the high-level business requirements, which also helps verify the scope.

Project Lifecycle

The project delivery took a waterfall approach for each ship, and once the design was completed, there was a long sequential construction phase (this would include procurement of materials). This was followed by the launch, fitting out (or outfitting), sea trials (or testing), and delivery. The first two ships would be built almost in parallel, delivered eleven months apart. Harland and Wolff would employ a rolling workforce, redeployed from ship to ship, which was the most effective use of workers' time and the most efficient use of the shipyard infrastructure, equipment, and tools, bringing economies of scale into the project. The project would use the best of industrial practices to build the two Olympic-class ships.

The project business case supported three ships, but the first two ships (*Olympic, Titanic*) had to be operational within a five-year period to fund the third ship (*Gigantic*, later known as *Britannic*).

Project Schedule

For all three ships, the design time was less than six months. For each ship, the construction phase was about three years, and then followed by the launch. The fitting-out phase would take another year. Finally, the sea trials were one–two months, and delivery and the maiden voyage took a single week. In all, the total elapsed time was about four years.

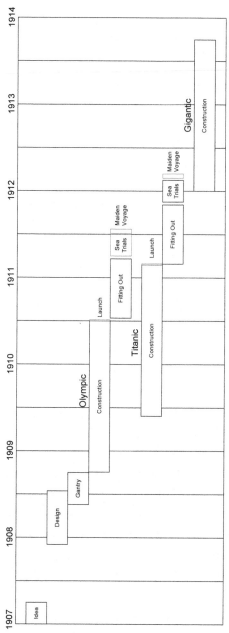

Figure 3.1. The project schedule for the three Olympic-class ships ran from 1907 to 1913. Each phase for a single ship was sequential (design, construction, fitting out, sea trials), and they could not be overlapped.

Design

The order to proceed with the design of both new liners was signed April 30, 1907. The introduction of iron and steel into shipbuilding accelerated the growth in ship size and scale. As a result, building techniques changed and required a more certain and unified process.

The Naval Design Process

Typically, this included variations in design such as preliminary, concept, contract, basic, and detailed designs. The design process translated the shipowner's requirements into the drawings, specifications, and other technical data necessary to build a ship. Naval architects led the process, with engineers (marine and production) and designers (structural) part of the core team. A team of design draftsmen (for steel, outfit, piping, and electric) and estimators supported them. As the designers and draftsmen began the drawings and calculations in parallel, the buyers started to source the equipment and materials. The design processes were iterative and subdivided into several phases during which the design was developed in increasing degrees of detail.

Figure 3.2. Naval architects, designers, and draftsmen preparing hundreds of drawings and plans at the drawing office at Harland and Wolff shipyards in Belfast. Note the equipment used on the tables.[1]

Figure 3.3. The spacious drawing office at Harland and Wolff in Belfast was purpose built, and it created an environment conducive to design. Note the high barrel ceiling with large skylights in the roof to maximize the natural light.[2]

Hull Design

The design of the ship's hull was critical because it determined the shape, size, and capacity of the ship. In the ensuing project, Harland and Wolff's naval architects had many design choices, and from the outset, they followed a project objective to maximize passenger comfort rather than passage speed. They never imagined breaking the Blue Riband record for crossing the Atlantic in record time. This adjustment meant the ship could be built with a broad U-shaped hull rather than a sleek and fast V-shaped hull (Figure 3.4), which increased the ship's volume by 23%, resulting in larger and more comfortable first- and second-class suites and cabins that greatly enhanced the passenger experience. In hull design, naval architects strove to find a balance between the hull's capacity, stability, and performance (Figure 3.5).

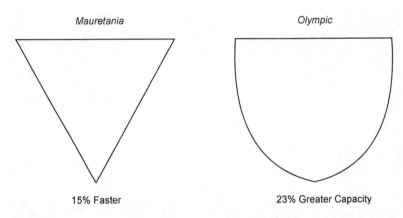

Figure 3.4. *Olympic-class ship's U-shaped hull (right) provided a greater carrying capacity than competing designs with a V-shaped hull (left).*

Design as a Competitive Differentiator

Likewise, today, you can copy your competition's approach or use emerging technologies to gain an advantage as in this story. If you try something different, such as exploiting a niche in the marketplace, you might achieve a far better economical payback. White Star's approach was well thought through as it was clearly differentiated from the competition's approach.

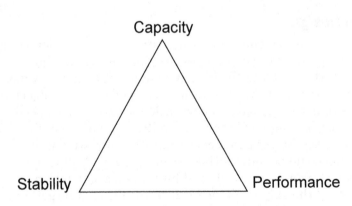

Figure 3.5: *Hull design for a naval architect is a careful balance between capacity, stability, and performance.*

Design Team

Harland and Wolff provided a design team that included
Alexander M. Carlisle and Thomas Andrews. Alexander
Carlisle was the head designer and responsible for
coordinating the designs. His chief area of responsibility
was the equipment used on the ships. Thomas Andrews, the
Managing Director of the Design Department, was responsible
for every drawing. Edward Wilding was Andrews' deputy and
responsible for all the design calculations.

Functional Requirements

As part of the project's design phase, the next step was to
identify, collect, and refine requirements as part of **Scope
Management**. The architects proceeded to transfer the
business requirements into the ship's functional requirements
and elaborate on these—what a system does, namely,
transportation and hospitality. The principal functions
included defining accommodation through cabins and suites,
catering, recreation, and entertainment. The focus was on
the luxurious splendor of the first and second classes and
raising the bar for the passenger experience. For example, a
spectacular first-class entrance stairway that went through
three decks, three elevators at the forward first-class
main entrance for conveying ten passengers, and a heated
swimming pool and Turkish baths. All this was awe inspiring
but technically very challenging for the architects.

A further aspect of design that could inspire awe was
with the funnels. The initial vision of three funnels per ship
was replaced by four to create symmetry and communicate
a sense of speed—more funnels equated with a faster ship, a
marketing stunt. It also reassured the third-class emigrants
that the ship was big and safe enough, an important part of
the **Communications Strategy** with the public, or external,
stakeholders.

Government Regulations and Requirements

The architects had to pay close attention to these with a ship crossing the Atlantic (3,200 miles). The ship had to be seaworthy, and machinery had to be within specific specifications. There were also many safety regulations to protect all on board. There was also a requirement for provisioning with adequate food, water, and fuel on board. The British Board of Trade, who visited the work site almost weekly, checked off the regulations throughout the project.

Likewise, most projects today transition through this phase smoothly. Functional requirements are tangible and readily understood by business users, executives, and the project team, and so are easier to define.

Nonfunctional Requirements

Investments also had to be made into nonfunctional requirements—everything that supports the functional requirements. The nonfunctional requirements are incredibly important because they define the operational characteristics of a system, or a system's delivery. For the project's architects, this included reviewing safety, performance, stability, security, maintainability, redundancy, and reliability. Therefore, no matter what the specifications for the functional requirements were, the nonfunctional requirements ensured the ship could deliver the functions for which it was designed.

Likewise, in today's projects, the nonfunctional requirements include availability (similar to safety), security and system management, scalability, portability, maintainability, and evolution. Similar to the project, the nonfunctional requirements ensure that the system delivers the functions for which it was designed.

Shipbuilder's Model

For the project, a 15-foot model, shown in Figure 3.6, was created. This was not just a showcase model used to promote the Olympic-class ships to prospective buyers. It exemplified also Harland and Wolff's and the project teams' capabilities. It also provided an in-depth view of the design. Finally, an accurate model could also be used for playing out what-if scenarios.

Figure 3.6. The shipbuilder's model for the Olympic-class used for risk identification and static testing during the design phase. The model is in the Liverpool Maritime museum.[3]

Risk Identification and What-if Scenarios

To determine the nonfunctional requirements, Harland and Wolff's architects used techniques like the shipbuilder's model to provide a quasi-simulated environment for determining how the logical and physical designs worked with the functionality of the ship. For example, was the power-to-weight ratio adequate, or how stable was the ship in rough seas? Regular transatlantic travel dated back 400 years. The architects were very aware of all the different dangers and risks to an ocean liner. Using the shipbuilder's model, "**what-if-scenarios**" for failure were modeled to determine alternate safety features and their optimum implementation. For example, scenarios for the ship running aground or scenarios involving front-end or side-impact collisions. The worst-case scenario imagined adjacent compartments flooded. This is where **risk identification** came in. The most common accident was a collision with another vessel, which would leave two compartments open to the sea.

Static Testing

Likewise, to define nonfunctional requirements today, many modeling techniques are helped typically by computer simulation. For example, one technique used for today's solutions is a flow analysis model that provides a high-level trace of each critical business transaction as it traverses the solution from end to end. Each component on the path is assessed for its nonfunctional characteristics. Another technique is to use this model for worst-case failure scenarios for single or aggregated parts along these paths, for example, the impact during a spike in transaction flows or during a cyber attack. This approach is also known as *static testing* or a dry run, or walk-through, as opposed to *dynamic testing*, which is done in the real world and not on paper.

Aside from the functional and nonfunctional requirements, which are also known as **product requirements**, **project requirements** address project management practices and guidelines. In today's projects, once the requirements are collected, they need to be managed, particularly for any changes. A **Requirements Traceability Matrix** records each requirement and tracks its attributes and changes throughout the project life cycle.

Selecting the Right Level of Safety

Harland and Wolff's architects had to assess the various emerging technologies and determine how these would deliver the various safety features (nonfunctional requirements). They had to make investment choices and effectively select a level based on the possible what-if failure scenarios. For example, the basic level was defined as having no safety features at all, where the ship had the characteristics of a closed rowing boat good enough to operate across the English Channel, but not in open sea. The highest level was defined as having a comprehensive set of safety features to include a full complement of 64 lifeboats; advanced water pumps that discharged vast quantities of water quickly (bilge and ballast pumps discharged 150 and 250 tons of water per hour, respectively); and advanced features such as a double hull (Figure 3.10), several (16) watertight compartments (Figure 3.11), 15 sealed bulkheads (Figure 3.12), and a system of pressurized air to contain water flooding; electric vertical-sliding watertight doors (Figure 3.13), a collision ram, submarine bells (Figure 3.14), automated control systems such as fog warning systems (timing whistle blasts in the fog to warn other ships), communications (phone and wireless telegraph or Marconigram (Figure 3.15)), and davits for the lifeboats (Figure 3.18). The ship had the characteristics of an oceangoing vessel good enough to make transatlantic crossings at full speed.

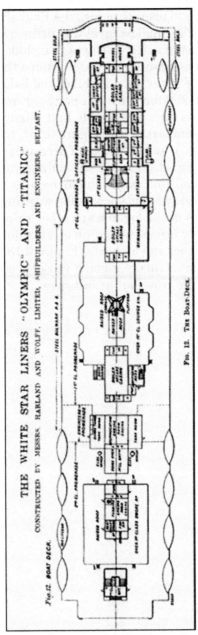

Figure 3.7. The Olympic-class's boat deck had 16 lifeboat stations (8 on each side) that could hold a full complement of lifeboats of 64 (quadruple stacked) or 48 (triple stacked).[4]

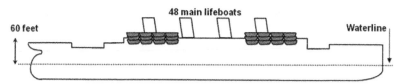

Figure 3.8. Side profile of the boat deck. The new Welin davits could hold up four stacked lifeboats across 16 lifeboat stations.

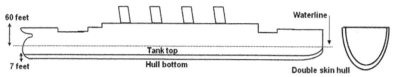

Figure 3.9. A double-hull design and bulkheads reduced the likelihood of a side or bottom penetration of the ship and flooding of the interior of the ship. The hull and inner bottom plating gave the ship its longitudinal strength.

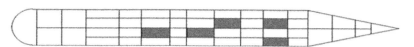

Figure 3.10. A double-hull bottom, and within it were many (44 of the total 73) watertight compartments that contained flooding in grounding situations. A system of pressurized air contained water flooding.

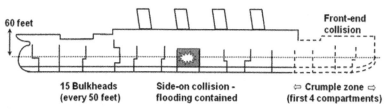

Figure 3.11. The ships had 16 sealed bulkheads that protected the ship from side-on collisions. The front end of the ship design was similar to a modern "crumple zone" in automobiles. It was designed to collapse on impact, but it protected the rest of the ship.

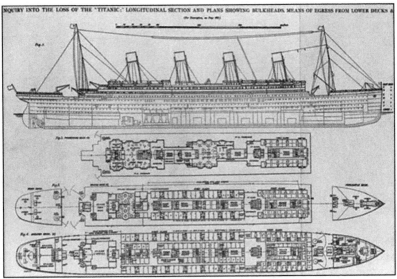

Figure 3.12. The ship was divided into 16 multiple watertight compart-
ments by 15 bulkheads to contain the flooding in case the double hull was
pierced in a collision. The bulkheads connected by angle irons to the decks
and inner bottom gave the ship traverse strength.[5]

Figure 3.13. Each watertight compartment had electric vertical-sliding watertight doors formed of heavy cast iron, and each was fitted with gears to control the speed of closing to thirty seconds. The doors were controlled from the bridge. It was an extremely important safety feature as it allowed rapid horizontal access through the length of the ship.[6]

Figure 3.14. Submarine signaling (bells) were used to detect underwater obstacles. They sounded under the sea. Two small boxes (in the image) were attached to each side of the bow inside the ship; inside was a water tank containing a microphone. From each tank, wires were run to a device, and the signal was heard by phone in the chart room.[7]

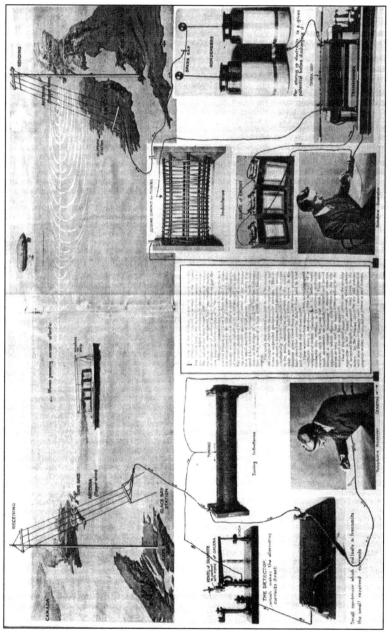

Figure 3.15: The wireless telegraph (Marconigram) used for ship-to-ship and ship-to-shore communications was one of the most important pieces of safety equipment on board.[8]

This approach is little different today where architects go through a similar process and assess how various emerging technologies will meet the nonfunctional requirements so they can decide how to make investment choices. There are many techniques and choices for improving the nonfunctional requirements of a solution and protecting it against potential problems by eliminating single failure points through replication and redundancy.

Perceived Safety of Iron Ships

When it came to safety, the public perceived iron ships to be very safe, going back to 1879 when *Arizona* (an early iron ship) collided with an iceberg going 15 knots in heavy fog and crushed 25 feet of her bow. However, after close inspection, it was found that the flooding was contained, and the ship continued the voyage. This made the headlines and changed the public perception of iron ships. Had *Arizona* been constructed from wood, she would have splintered to pieces. As a result, there was a sharp rise in ticket sales.

Figure 3.16. In 1879, Arizona (an early iron ship) collided with ice and saw a sharp rise in ticket sales as iron ships were seen to be safer than wooden ships. She had seven transverse (vertical) bulkheads up to the top deck.[9]

Rolling Wave Planning

In today's projects, we apply this progressive elaboration technique to address uncertainty in the project's future work, that is, longer term deliverables are identified at a high level and decomposed as the project progresses. This is usually the case when emerging technologies are used.

Finalizing the Design

In the end, Harland and Wolff's architects opted to go with the highest level of safety and incorporate all the latest and advanced safety technologies. After all, they were building the best ship they could, based on the latest technologies, and they had the budget to do so. Even though many emerging technologies for safety had been used in other ships, this was the first time that all were put into a whole package and proclaimed in White Star's marketing literature. Even White Star's Captain Smith became an important spokesperson, proclaiming after his trip across the Atlantic on the *Adriatic:*

"I cannot imagine any condition which would cause a ship to flounder...Modern shipping has gone beyond that."

—Captain Smith

Figure 3.17. Captain E. J. Smith of Olympic and Titanic.[10]

Project Phase Gate

As the project was ready to move into construction, the architects had to confirm their detailed design with the White Star board and Director Bruce Ismay. This was a major phase gate for the project.

Steering Committee Meeting

This was a meeting between President Bruce Ismay and Harland and Wolff's architects, who presented to him the various safety features, which included the vision for the lifeboats. Sixteen stations would house a pair of davits where each would carry quadruple lifeboats (64 in all). Carlisle had suggested a new type of larger "Welin" davit (Figure 3.18) that could carry more lifeboats, up to four. Bruce Ismay, whose focus was to create the ultimate passenger (first-class) experience, asked a simple question about which deck would house these lifeboats. The architect's intent was to house the lifeboats on the highest deck of the ship known as the boat deck, a deck used by first-class passengers for strolls and walks. Bruce Ismay recognized immediately that a wall of quadruple stacked lifeboats would obscure the ocean vistas from this first-class deck. Logic reasoned that passengers did not like lifeboats on their decks and preferred the open space to walk around the decks. Besides seeing so many lifeboats could make passengers feel uneasy and affect their overall experience.

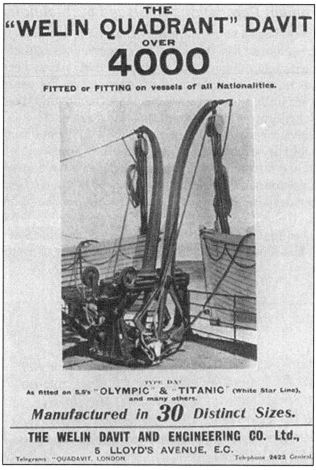

Figure 3.18. The Welin Davit Company proudly announces that its davits are carried on the Olympic-class ships; the association provided good exposure for the company. A pair of Welin davits carried up to four lifeboats.[11]

Compromise One—Number of Lifeboats

The architects were presented this dilemma. After some thought, they compromised and agreed to a double and lower bank of 32 lifeboats. Each pair of davits would carry double stacked lifeboats, with a seating capacity of 1,960, more than the regulations on lifeboats that stipulated a seating capacity for 962 people.

Government Regulations Replaced by Technology

The Olympic-class ships were fully compliant with the lifeboat regulations of the day. The government regulations on lifeboats had not been updated since 1896, partly because advances in technology deemed it unnecessary. For example, ships were built with watertight compartments and wireless technology. Ships also used well-travelled sea routes, so help was always close at hand. The provision of lifeboats was a matter for the shipowners to consider. To his credit, Alexander Carlisle had misgivings about the regulations written when the largest tonnage for ships was about 10,000 tons. German and US rules already required a greater proportion of lifeboats than the United Kingdom did. The British Board of Trade's antiquated lifeboat regulations were based on cubic feet of lifeboat space per ton of ship, not on the number of people aboard.

Figure 3.19. The Henderson collapsible boats on the White Star liner Arabic (swinging out). Four were used on both Olympic and Titanic, taking the total to 20 lifeboats on each ship, fewer than one-third of the originally specified lifeboats (64).[12]

Compromise Two—First-Class Dining-Room Saloon

The second compromise related to the required space to entertain the first-class passengers and maximize the customer experience. Bruce Ismay wanted a spacious first-class dining-room saloon (114' x 92') and reception room (54' x 92') (against the ship length and beam of 882' x 92'), the largest room ever to go to sea, and to sit 532 people. It was central to a variety of balls and gala dinners to make the voyage an unforgettable experience. To house this public area on Deck D, three of the horizontal bulkhead walls had to be compromised in height so they did not cut through it.[13] These bulkheads (just more than 50' apart) should have reached the top deck but were only 10' above the water line. This, interestingly, still fell within the Board of Trade regulations.

Figure 3.20. The first-class dining room, the largest room ever to go to sea, was a centerpiece for the first-class passage. At 10,488 sq ft (114 ft x 92 ft), it compromised three vertical bulkhead walls in height, which ended at the floor of Deck D.[14]

Figure 3.21. First-Class Dining-Room Saloon was 114 ft x 92 ft, and the Reception Room was 54 ft x 92 ft, a combined 168 ft against the ship length of 882 ft, around 20% of the length, and affecting three bulkheads.

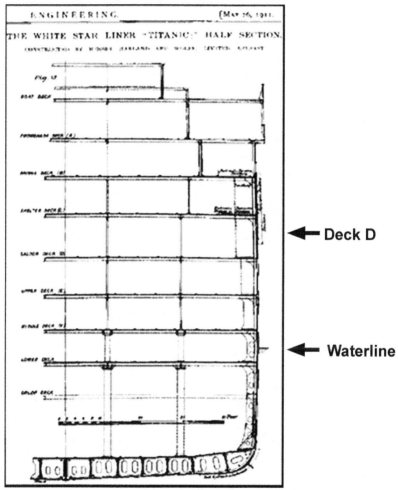

Figure 3.22. Shows the height of Deck D where the first-class dining room sits two decks above the water line (lower deck). The bulkheads ended below Deck D. Deck C (above D) was the highest deck, which extended continuously (no bulkheads) from bow to stern.[15]

Compromise Three—Double Hull

The third compromise related to the double hull, which initially was planned along the bottom to continue up the sides, providing a double skin. However, it was extremely expensive and ate much space. Where the double skin was 7 feet wide, this would take 15 feet away from a 92-foot beam. Across 880 feet, this was significant and was counter to the project mantra of maximizing passenger space. The compromise limited the double hull just to the bottom, 7 feet deep, and well below the water line.

Compromising the Nonfunctional Requirements

Notably, Bruce Ismay, who wanted to create the ultimate passenger (first-class) experience, pushed Harland and Wolff's architects hard to make compromises to the safety features. One by one, compromises were made in the nonfunctional requirements as the focus stayed on the functional requirements. The architects had the option to push back but caved in under this executive business pressure. A complacent view had evolved among the project team that the Olympic-class ship was a huge lifeboat, so the compromises seemed less severe. It is significant that during this phase, between four and five hours were devoted to discussing décor and fittings, but no more than ten minutes were given to lifeboat capacity.

Tracking Micro-decisions

Similarly, in today's projects, hundreds of micro-decisions are made daily or weekly, decisions deemed too technical for business executives. Nonfunctional requirements are typically beyond the comfort zone of most business executives, so any compromises to these might not be readily understood, so this type of "corner cutting" is less noticeable. Yet, it might have a major impact later, once the solution is in production and can affect the business itself.

Board of Trade Approvals

The overall dimensions and form of the ship were agreed,
and the preliminary design and general arrangement of the
ship (quarters, spaces, and compartments) were approved.
Before construction was started, structural plans were
prepared by Harland and Wolff and submitted to the Board
of Trade for approval on June 3, 1908. These included
general specifications of the hull and the sizes of all steel,
iron, and wood forming the structural parts of the vessel.
All these features were to be distinctly marked according to
the requirements of the Board of Trade. The plans included
the construction of the watertight bulkheads, designed in
agreement with the latest report of the Board of Trade's
Bulkhead Committee.

Final Design Review Meeting

A party of distinguished guests arrived at Harland and Wolff
in Belfast on July 29, 1908. They had come from the White
Star Line's main office to examine the shipbuilder's concept
plan for the huge new ships that had recently been discussed
between Pirrie and Bruce Ismay. The party included, among
others, Bruce Ismay, Harold Sanderson, and other White Star
directors. They examined the scale drawings prepared by
Harland and Wolff draftsmen under the direct supervision of
Pirrie, assisted by his nephew Alexander Carlisle, the head
designer.

Procurement Management

The structural plans were returned to Harland and Wolff who
then ordered the steel and iron for the project. The volume of
materials required for constructing the ships was immense,
for example, 24,000 tons of steel went into each ship, and each
ship required 3 million rivets. Harland and Wolff purchasing
procedures had standing agreements with pre-qualified
suppliers. These suppliers had been selected based on the
needs of various projects, and many qualification criteria were
used, including price, life cycle cost, alignment to the project,
technical abilities, warranty, past performance, current

working relationship, financial capacity, capacity to produce, reputation, size, and type of business. These suppliers had to be alerted to the project, its demands, its schedule for delivery, and lead time criteria.

Conduct Procurements

This process involved obtaining responses from these suppliers. In the case of the project, several suppliers were favored but from a preferred supplier list, for example, rivet suppliers. Through the project, Harland and Wolff procured a steady stream of equipment and materials—everything from rivets to complex steam turbines. Harland and Wolff was prepared for the volume of the required procurements, which was the reality of the business. An examination of records from the National Archives in the United Kingdom reflects the project scale. It outlines the delivery to the shipyards of batches of materials, various parts, and prebuilt sections. Weekly, the Board of Trade inspectors were on hand and closely inspected the latest delivery.

Contract Signed

The draftsmen proceeded to prepare the detailed structural plans for the workers in the shipyard. On July 31, 1908, the final designs of the two ships were completed, and an agreement was signed to start construction. Bruce Ismay would approve any changes. In 1908, White Star Line issued £2.5 million of additional shares to cover the project costs (with 1909 profits at about £1 million).

Chapter Wrap-up

Conclusion

As in many projects today, the design went smoothly until it went through a stage gate, where a struggle took place within the project team where the success of the business strategy overrode other considerations. The project team made the mistake of believing the initial design assumptions (the Olympic-class ships were huge lifeboats) and not testing these far enough. In short, the people "who should have gotten it"—the architects—allowed the compromises to pass. This is a good example of decisions related to esthetic factors compromising the individual safety features.

Key Lessons for Today

Today, many projects are compromised in the design phase, almost innocuously. But the impact of these compromises might not be apparent, as any problems might not surface until days, months, or even years after the project is completed, and the solution is in production. Project managers need to ensure that, as part of scope management, as much attention is paid to the nonfunctional as the functional requirements.

Very often, the nonfunctional requirements are sacrificed because they are less visible, and their importance is not highlighted to business people, project leaders, or executive decision-makers. With hundreds of micro-decisions made weekly, project managers also need to aggregate the potential impact of these decisions on requirements. This impact needs to be made clear to the business leaders and executives.

Today, typically before project managers complete this phase, they would do the following:

- Ensure that due diligence is paid by the project team to granular decisions required. The project manager needs to pay attention to these, aggregate them, and determine any risks.

- Ensure that the steering committee is fully engaged in the project and understands its role, decisions being made.

- Bring in users to define the requirements (business, functional); avoid second-guessing them through other representatives.

- Determine the integration requirements between the proposed solution and the existing services (environment) and any dependencies.

- Avoid complexity and strive for simplicity in the design.

- Design to support the different types of users. In addition, design "learnings" into tools where experienced users might not have time for training.

- Establish the service-level targets to guide the architect(s) through the design. Service levels captured later in the project can be troublesome and risky.

- Avoid acquiring solutions driven by an emerging technology. Ensure due diligence by going through a fit-gap analysis, examining the potential costs and assessing the risks.

- Avoid under-investing in nonfunctional requirements. A mistake in these is far more costly than in the functional requirements, which can always be added in future iterations.

- Avoid one technology; lack of diversity increases susceptibility to problems.

- Define an end-to-end view of the solution, for example, along all critical transaction paths.

- Walk through the design early to catch potential problems.

- Review government regulations that could affect the solution, assess whether they are keeping up with the technology, and then plan accordingly.

- Ensure expectations are understood clearly if any compromises are made.

Educators

Discussion points:

- Discuss the techniques that would be considered today, such as future scenario planning.

- What alternatives could the architects have taken when pushed into making compromises to the nonfunctional requirements?

- What was the role of the executive sponsor Bruce Ismay? Was he merely looking for opportunities in the project? Was he paying attention to the risks?

- Was the executive sponsor Bruce Ismay overstepping his boundaries? Or was he just looking after the core mantra of the project, the ultimate customer experience.

- Discuss the changing risk profile with each compromise.

- Was there a danger that emerging technologies could outpace the project?

- How well was the project phase completed?

CHAPTER 4

Construction Phase

In This Time Frame

○ September 17, 1908—Harland and Wolff given orders to proceed with construction of *Olympic* (yard number 400) and *Titanic* (yard number 401).

○ December 1908—*Olympic*'s keel laid.

○ March 1909—*Titanic*'s keel laid.

○ October 1910—*Olympic* launched.

○ October 1910—*Titanic* fully plated.

○ May 2, 1911—*Olympic*'s basin and sea trials.

○ May 31, 1911—*Titanic* launched.

Overview

This chapter looks at how the construction phase unfolded, and as the Olympic-class ships rose in height, the perception of invincibility continued to grow among the project team. The marketing plan reinforced this perception with the public.

Mold Loft

Work started in the Mold Loft where the first fairing of the lines done by the draftsman was turned over to the lofts-men. The Mold Loft was a large floor (several 100 feet long and one hundred feet wide) on which the lofts-men chalked the lines of the cross sections of a ship at full size and the length at quarter scale. The work included making templates, or "molds," of heavy paper or thin wooden boards for all the ship's structural parts. A template for a steel plate consisted of a full-sized pattern of the plate marked out and showing in detail all punched or countersunk holes, scarves, bends, and angle lines.

Figure 4.1. Laying of the ship's lines in the Mold Loft at Harland and Wolff shipyards. This process transferred the design from the drawing office to full-scale physical templates. It was used well into the twentieth century until replaced by Computer Aided Design.[1]

Human Resources Management

Today, regarding **Human Resources Management**, we typically **acquire a project team**, which is then **developed and managed**. For Harland and Wolff, this was straightforward in 1908, as the company had a standing 14,000 workers to build up to nine ships at any one time. Harland and Wolff was the largest employer in Belfast where 3,000 workers worked on Olympic-class project at any time.

System of Trades

The Harland and Wolff workforce was organized by a system of trades underpinned by an apprenticeship system. Since the mid-nineteenth century, the system combined work-based training (on-the-job training) and attendance at a local technical institute (off-the-job training). In a trade, an apprentice graduated to a journeyman (a tradesman or craftsman), and then, possibly a master. The trades were tightly knit, and they came from closely knit communities (in the past, guild members had lodged together, a form of co-location). There was an emphasis on the **training and development** of members into a cohesive project team, working with a set of ground rules. The team members were interdependent, committed to working together, and as a whole, the team was accountable.

Figure 4.2. Shipwrights or boat builder apprentices went through an extensive seven-year apprenticeship (four to five years was the norm) that required attending apprentice school at the Harland and Wolff shipyard and working in the various shops, molding loft, and on ships (see Appendix and Thomas Andrews apprenticeship).[2]

Organization of Trades

The elite trade in shipbuilding was the shipwrights (boat builders with a seven-year apprenticeship). The skilled trades (50%–60% of the workforce) included the platers, riveters, drillers, pattern makers, fitters, founders, smiths, boilermakers, electricians, carpenters, cabinetmakers, and artisans. There was also a large force of trade laborers. The most job security was for the Established-men who were paid less (24 shillings per week) than the Hired-men (25 shillings per week) who had little security. Although it was easier to **manage a project workforce** with a level of insecurity, by all accounts, there was a good relationship with the Harland and Wolff management and mutual respect between the two. The workforce was simultaneously working on six other liners, including *Demosthenes* and *Galway Castle*, and two White Star tenders (*Nomadic* and *Traffic*) for use at Cherbourg. Three thousand workers worked on the project at any time, working 49 hours per week for 50 weeks a year.

Construction Under Way

In 1908, *Engineering* (Magazine) described the world-class construction project under way at Harland and Wolff, which included a power plant setup and new gantry being built for *Olympic* and *Titanic*. Both ships were built using the same plans, but *Titanic*'s **construction schedule** lagged seven months behind *Olympic*. The tested approach of shipbuilding laid a keel of a "flat-plate" design formed by a single thickness of plating. On to this, a "vertical keel" was stood and attached, which was then formed into a double-bottom space divided into rectangular cells by the floors and longitudinal.

Figure 4.3. The laying of Titanic's keel was a tested approach in the construction project practiced for thousands of years. Note the cellular double bottom with the double-bottom space divided into rectangular cells by the floors and longitudinals.[3]

Project Work Breakdown Structure (WBS) Level 2

In today's projects, the first WBS (part of **Scope Management**) is further **decomposed** to create a set of deliverables for construction. This creates small **work packages** (the lowest level of the WBS) that are more accurately estimated and more easily assigned and tracked.

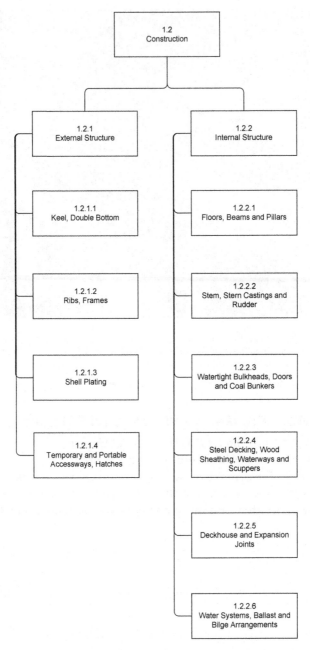

Figure 4.4. Projection of what a possible project work breakdown structure would have looked like for the construction (level 2).

Cost and Time Management

During the construction phase, much attention was paid to optimizing the project costs and schedule. For example, automated riveting was introduced to reduce labor costs and shorten the critical path.

Optimizing Project Costs and Schedule

As the project progressed, when it came to **Cost Management,** Harland and Wolff looked for opportunities to optimize costs. One area was riveting, which traditionally was done by hand. It was a laborious process of hand hammering a three-inch red-hot slug between two plates. Automated riveting was introduced to reduce labor costs but only along straight sections of the hull. There were more than three million rivets, so plating the ship was slow. The riveting not only created watertight seams but also strengthened the ship's integral structure, as the number of rivets was increased in some areas such as in the tank top (top of the double hull).

Determining the Activity Dependencies

When it came to **Time Management**, for the most part, the activities were sequential with little opportunity to save time by running activities in parallel. These **mandatory dependencies** required a natural order. For example, the keel preceded the double bottom, which, in turn, preceded the ribs and frames. Today, these relationships are called hard logic.

It is now common to see more flexibility in the sequencing where **discretionary dependencies** allow activities, whether best practices, project conditions, or external events, to occur in a specific order. These relationships are called soft logic, or preferred logic.

The project had to consider any **external dependencies** that could affect the project, namely, ensuring there was always an adequate stockpile of materials. This was a role for the buyers coordinating the suppliers shipping vast numbers of materials (steel plates, rivets) almost continuously, or vendors and their equipment.

Optimizing Project Schedule

The work along the **critical path**, the riveting of steel plates, was arduous and difficult and the longest activity in the project. Rivet squads of three men and two boys completed it. One boy heated the rivets, and the other inserted them in the hole. This material was more malleable than steel. The rivet was struck a few blows to lay it up, so it would bed fairly on the plate. It was then "held up" with the hammer and struck either side in rapid succession.

From a **Time Management** perspective, it was difficult to find areas were the schedule could be shortened. The introduction of hydraulic riveting machines accelerated the riveting process and helped shorten the schedule, but this was somewhat limited by where the machines could be used, as they were restricted by moderate bends in the plates.

Figure 4.5. Rivet Furnace from Rockwell shows how rivets were heated on deck. Riveting was labor intensive for the workforce.[4]

Confidence Grows in the Project

As the construction phase unfolded (Figure 4.4), there was no comprehension that anything was seriously wrong with the ship and its design. In many ways, it was almost too late to do anything about it. Even though the ship's nonfunctional requirements had been compromised, the belief persisted among Harland and Wolff's architects that the Olympic-class ship was practically unsinkable. The project **communication plan** was not altered. The broad hull design, the sheer size of the ship (40% longer than the last great ships built), and the aggregated effect of all the advanced safety features and technologies would protect the ship from whatever nature handed out.[5]

Justifying the Number of Lifeboats

In this atmosphere, it is easy to see how the architects justified limiting the number of lifeboats to 32, the most serious compromise of all. Lifeboats were now deemed an added safety feature, something useful if the Olympic-class ship had to rescue passengers from other ships in distress. The Olympic-class itself was never imagined to be a ship in distress.

Figure 4.6. Progress in constructing the 10 decks, which outlines the beam heights of the decks, pillars, web frames, and other structural members. Note the workers at the center.[6]

At this point in the project, the perception among the project team was that the Olympic-class was invincible. It could survive any situation, so greater risks could be taken. Bruce Ismay further propagated the perception that the ship was practically unsinkable when he realized it was another feature he could use as part of White Star's marketing of the ships.

Quality Management

This was very important during the construction phase and manifested itself through the **quality standards** set up with suppliers and contractors for materials and equipment; **quality assurance** with the thousands of visits made by Board of Trade inspectors; and the **quality control** through all the continuous inspections that were an important part of this phase.

Quality Standards

Harland and Wolff partnered with a number of iron and steel suppliers in Scotland and England for rivets and steel plates. Batches of 10,000 to 25,000 steel plates were ordered. There were **quality standards** in the steel's strength and ductility. The Board of Trade required standard tests for all steel on boilers, bulkheads, and watertight doors. Harland and Wolff always ordered best-quality stock that was already tested according to the day's standards.[7]

Figure 4.7. Olympic and Titanic towering in the background. The stockyard at the forefront holds the thousands of steel plates and other materials. Note the worker (bottom) standing by the plates, each measuring 6 feet by 30 feet.[8]

Quality Assurance

As part of the overall process, Board of Trade inspectors completed several thousand visits to the project. During these visits, inspectors were concerned about standards being met and laws and regulations being followed. Quality control tests provided **quality metrics** to review, for example, stress tests and tests for the tensile strength of the steel supplied as measurements to the project. This is similar to today's world where quality audits are independent evaluations of quality performance to ensure that products are safe and fit for use, and improvement opportunities are identified. The Board of Trade inspector visits culminated in the sea trials, which, when successfully completed, would lead to the final acceptance of deliverables.

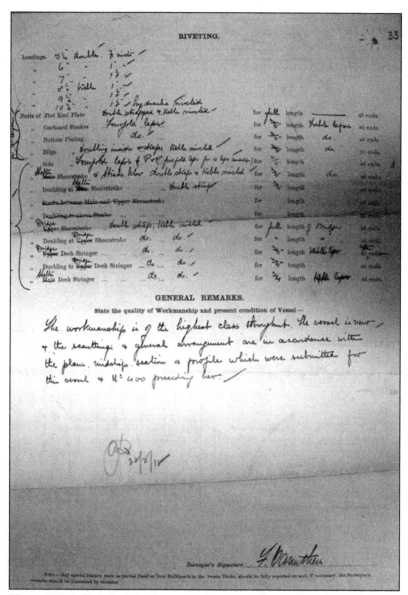

Figure 4.8: A Board of Trade report that states the quality of the workmanship for riveting: "the workmanship is of the highest class throughout," signed by Francis Carruthers.[9]

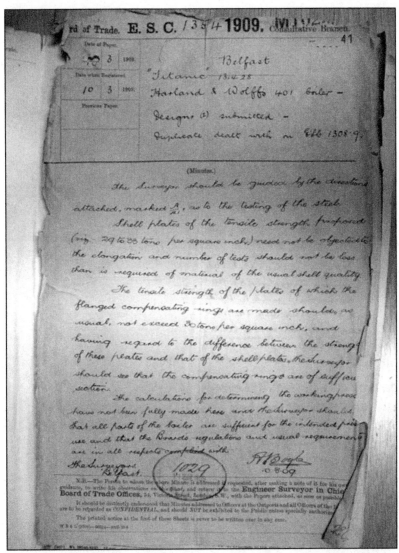

Figure 4.9. The Board of Trade minutes outline the number of tests required to test the steel's tensile strength. The last paragraph states, "all parts of the boiler are sufficient for the intended pressure and the Boards regulations and usual requirements are in all respects complied with."[10]

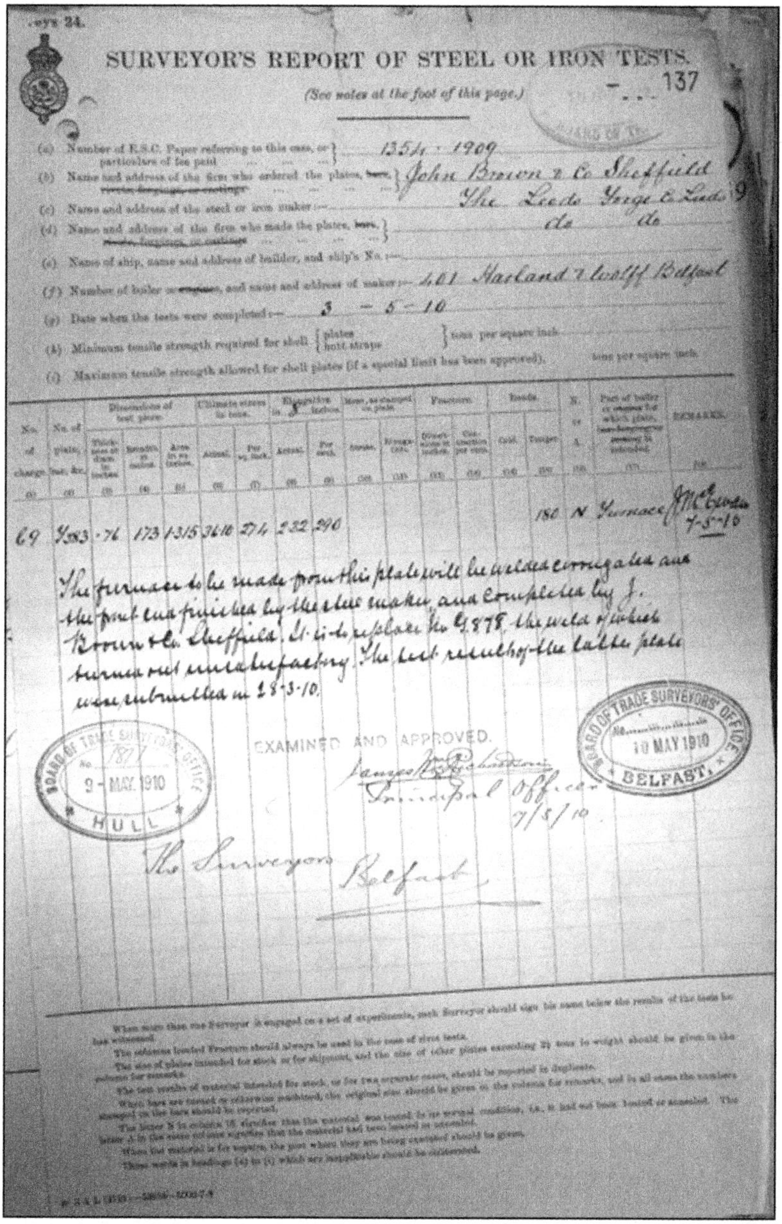

SURVEYOR'S REPORT OF STEEL OR IRON TESTS.

(See notes at the foot of this page.)

137

(a) Number of R.S.C. Paper referring to this case, or particulars of fee paid: 1354 - 1909

(b) Name and address of the firm who ordered the plates, bars, rivets, forgings, or castings: John Brown & Co Sheffield

(c) Name and address of the steel or iron maker: The Leeds Forge & Leeds

(d) Name and address of the firm who made the plates, bars, rivets, forgings, or castings: do do

(e) Name of ship, name and address of builder, and ship's No.:

(f) Number of boiler ..., and name and address of maker: 401 Harland & Wolff Belfast

(g) Date when the tests were completed: 3 - 5 - 10

(h) Minimum tensile strength required for shell {plates / butt straps} tons per square inch

(i) Maximum tensile strength allowed for shell plates (if a special limit has been approved), tons per square inch

69 | 9383 | ·76 | 173 | 1·315 | 36·10 | 27·4 | 2·32 | 29·0 | 180 | N | Furnace | J McEwen 7-5-10

The furnace to be made from this plate will be welded corrugated and the part end finished by the steel maker and completed by J. Brown & Co Sheffield. It is to replace No 9878, the weld of which turned out unsatisfactory. The test results of the latter plate were submitted on 18-3-10.

EXAMINED AND APPROVED.

Principal Officer 7/5/10

BOARD OF TRADE SURVEYORS' OFFICE No. 1877 9 MAY 1910 HULL

BOARD OF TRADE SURVEYORS' OFFICE 10 MAY 1910 BELFAST

The Surveyors Belfast

Figure 4.10. This paragraph states the furnace is made from this plate (delivered by John Brown of Sheffield), and it replaces one where the weld was found unsatisfactory (failed its tests). The test results were submitted 18-3-10. This highlights the corrective actions process used.[11]

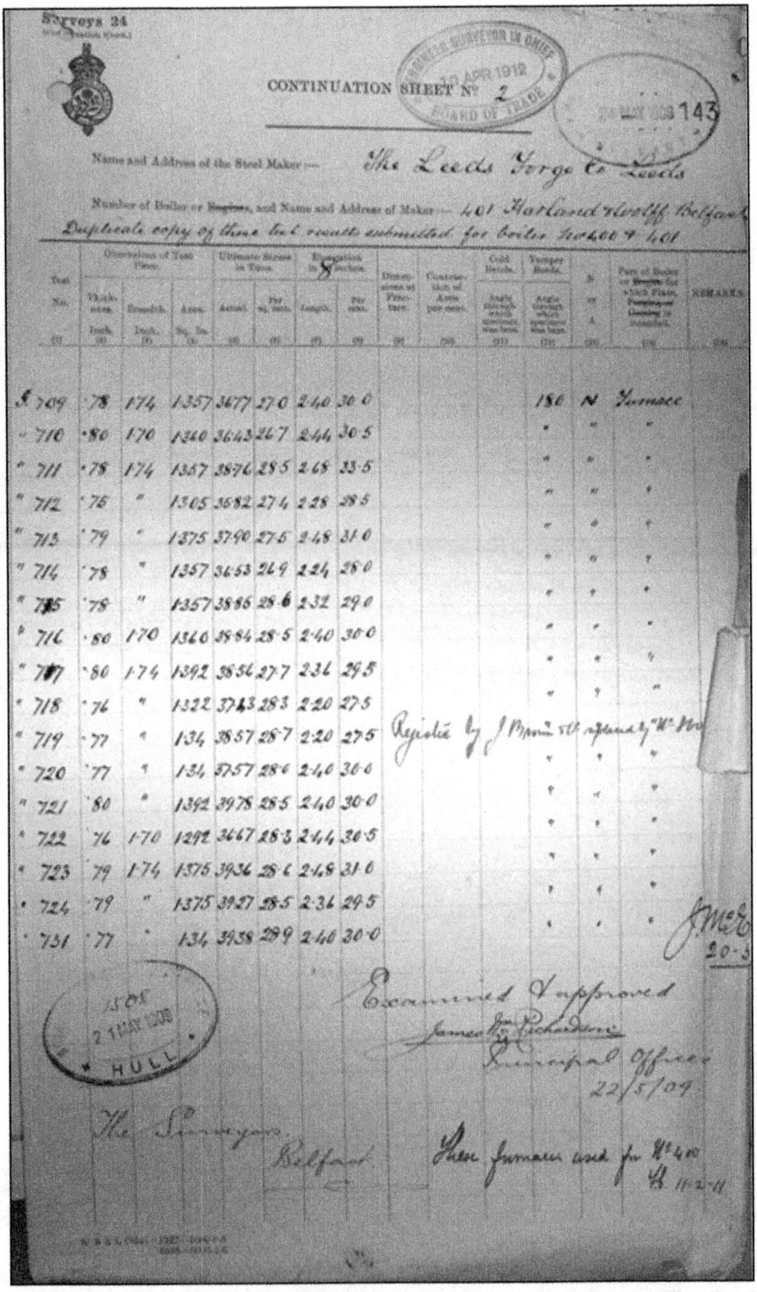

Surveys 24

CONTINUATION SHEET No 2

Name and Address of the Steel Maker — The Leeds Forge Co Leeds

Number of Boiler or Engines, and Name and Address of Maker — 401 Harland & Wolff Belfast

Duplicate copy of these test results submitted for boiler Nos 400 & 401

Test No.	Thickness (Inch.)	Breadth (Inch.)	Area (Sq. In.)	Ultimate Stress in Tons: Actual	Ultimate Stress in Tons: Per sq. inch	Elongation in 8 inches: Length	Elongation in 8 inches: Per cent.	Dimensions of Fracture	Contraction of Area per cent.	Cold Bends	Temper Bends	N or A	Part of Boiler	Remarks
S 709	·78	1·74	1·357	36·77	27·0	2·40	30·0				180	N	Furnace	
" 710	·80	1·70	1·360	36·43	26·7	2·44	30·5				"	"	"	
" 711	·78	1·74	1·357	38·76	28·5	2·68	33·5				"	"	"	
" 712	·75	"	1·305	36·82	27·4	2·28	28·5				"	"	"	
" 713	·79	"	1·375	37·90	27·5	2·48	31·0				"	"	"	
" 714	·78	"	1·357	36·53	26·9	2·24	28·0				"	"	"	
" 715	·78	"	1·357	38·85	28·6	2·32	29·0				"	"	"	
" 716	·80	1·70	1·360	38·84	28·5	2·40	30·0				"	"	"	
" 717	·80	1·74	1·392	38·56	27·7	2·36	29·5				"	"	"	
" 718	·76	"	1·322	37·43	28·3	2·20	27·5				"	"	"	
" 719	·77	"	1·34	38·57	28·7	2·20	27·5	Rejected by J Brown & replaced by No ...						
" 720	·77	"	1·34	37·57	28·6	2·40	30·0				"	"	"	
" 721	·80	"	1·392	39·78	28·5	2·40	30·0				"	"	"	
" 722	·76	1·70	1·292	36·67	28·3	2·44	30·5				"	"	"	
" 723	·79	1·74	1·375	39·36	28·6	2·48	31·0				"	"	"	
" 724	·79	"	1·375	39·27	28·5	2·36	29·5				"	"	"	
" 731	·77	"	1·34	39·38	28·9	2·40	30·0				"	"	"	

J. McE 20-5

Examined & approved
James Wm Richardson
Principal Officer
22/5/09

The Surveyors,
Belfast

These furnaces used for No 400
11-2-11

HULL

Figure 4.11: This report shows the stress testing undertaken by The Leeds Forge Co. for steel for the boiler. Note J. Brown rejected Test 719.[12]

The boilers had to withstand significant pressures. The steel for the boilers tested was tested by the manufacturers and then inspected by the Board of Trade (Figure 4.11).

Figure 4.12: Boilers laid out, arranged, and ready for installation in the Fitting-Out phase. Note the boiler-fitter at the center.[13]

Quality Control

Inspections were an important part of this phase and completed within each trade, as required. For example, rivet counters were not only there for testing for loosely fitting rivets, but also for taking corrective actions, if needed. Several methods were used. Typically, rivet counters would tap the rivets for any that were loose fitting. When found, these were marked as failing the test and not given a seal of approval, indicating they needed **corrective actions**. They were never caulked, but renewed and re-riveted, which was done when detected by the rivet squads attached to the rivet counter. As the construction neared completion, a secondary method was added where the ship's smaller areas were flooded on the inside to check for leaks on the outside. A third method was to insert bolts, instead of rivets, which would be tightened to detect loose rivets.[14] Rivet counters also counted the number of rivets completed by each rivet squad. These men were not very popular with the workforce and had to have an extra incentive to do this work, which was an extra £20 a year.

Project Mortality

The shipbuilding industry was dangerous and had a high mortality rate. Callous as it might seem, in the period of this project, the accepted death rate was one man per £100,000 invested. The 1906 Compensation Act ensured payouts around £300 for a worker, or about 2½ years of a worker's salary. Through the project, there were eight deaths where falling from heights made up the highest percentage.

Further Compromise on the Number of Lifeboats

Before the completion of the construction phase, a "gate meeting" for the project was held. The architects had agreed to downsize the number of lifeboats to 32 from Alexander Carlisle's original plan that incorporated 64 lifeboats. Bruce Ismay, who now solely represented the White Star board, wanted a further downsize to one stack of 16 lifeboats because it provided more space for verandas, sunny decks, and sports, and it was still above the Board of Trade regulations. Again, the architects compromised to Bruce Ismay's request. The reasoning behind the decision was that the aggregated safety features still put the ship way before any other ship. The view was that the ship was a lifeboat in itself, and the only need for lifeboats was to save passengers on other ships in distress. With a sense of guilt, four Englehardt collapsible boats were added, so the overall lifeboat capacity could accommodate 1,300, about 33% of all on board (maximum capacity passage). This was still more than the regulations on lifeboats stipulating a seating capacity for 962 people, or 25% above the regulations.

Figure 4.13. A vision of the boat deck for the first-class passengers with its open space and uninterrupted ocean vistas because of its single line of lifeboats. Bruce Ismay perceived this as critical to the passenger experience.[15]

Figure 4.14. A February 2, 1910, editorial cartoon in Puck titled The Central Bank—Why should Uncle Sam establish one, when Uncle Pierpont is already on the job? J. P. Morgan was caught in a crisis, and he was deflected from being closely involved with the White Star project.[16]

Alexander Carlisle Resigns

On June 30, 1910, Alexander Carlisle, age 56, found working with Lord Pirrie (his brother-in-law) difficult, and he resigned. Thomas Andrews took over. This was a severe blow to the project credibility. Carlisle went on to work for the Welin Davit Company, whom he had recommended in the design phase. The concern over the number of lifeboats was likely very influential in his resignation.

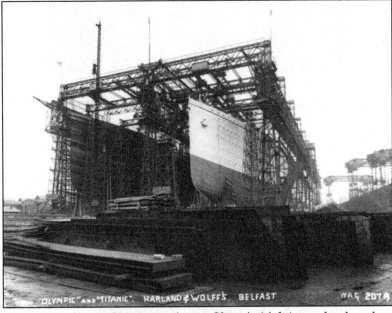

Figure 4.15. Shows the construction on Olympic (right) completed, and as she is painted in her light colors just before her launch. Note in the foreground are the remaining steel plates (6 by 30 feet) of the stockyard required to complete Titanic.[17]

Scope Management

As a project proceeds through the construction phase, one very important activity is verifying the project scope. Scope verification occurs as important deliverables are created or at the end of each project phase.

Scope Verification

The ship's launch was an end of project phase checkpoint (construction to fitting out), and it was used to verify that the deliverables the project created aligned with the project scope. Scope verification and quality control are closely related. The work's quality contributes to scope verification, where poor quality results in a failure in scope verification. A very large public crowd, dignitaries, and officials, who would put the ship under a level of scrutiny, would see the launch. Therefore, the ship had to be closely inspected for verification and, in turn, acceptance.

Protecting the Project Scope from Change

Change control ensures changes are agreed, and it determines whether a scope change has happened, and then manages the scope changes when they happen. The question remains why the change control mechanism in place didn't prevent downsizing the number of lifeboats to 16.

Olympics Spectacular Launch

As part of **Communications Management**, the outbound communications were sustained through the project that built up the brand through an advertising campaign of posters and interviews given to the media for articles. Unsurprisingly, there was much interest in what was probably the largest project in the world. *Olympic*'s launch in October 1910 attracted a massive 100,000 crowd of people, about a third of the city's population. As launch time approached, the stands were filled, and the banks of the Lagan River were lined with spectators, which highlights how successful the campaign had been.

For the launch, Bruce Ismay had *Olympic* painted a light gray and ochre, a striking contrast to maximize the launch's impact. It also made the ship far clearer in black-and-white photographs of the day (Figure 4.16). Her hull was repainted after the launch.

Figure 4.16. Olympic's launch, as she was released into the River Lagan in her light gray and ochre colors, a striking contrast to maximize the impact. The ship was merely an empty shell. All significant equipment such as the engines, boilers, and control systems still had to be installed.[18]

Figure 4.17. Another example of media coverage through the renowned Scientific American magazine (November 1910). This front cover proudly captions the image as "THE LAUNCH OF THE 50,000 TON OLYMPIC, THE LARGEST SHIP IN THE WORLD." This was exactly the publicity Bruce Ismay sought.[19]

Olympic *Fitted* Out

Olympic was launched and towed to the outfitting wharf
with nearly all the steelwork complete. This final phase of
the project would take nearly a year. There was much work
to do to turn *Olympic* from an empty shell of the vessel into
a floating palace, which included installing engines, boilers,
much of the important equipment, and control systems. All the
upper superstructures, such as the bridge and wheel house,
funnels, masts, and lifeboats, were also missing.

Project Work Breakdown Structure (WBS) Level 2

In today's projects, it is important to develop a WBS for all
project parts by further **decomposing** it and creating a
set of deliverables for fitting out, which creates small **work
packages** (the lowest level of the WBS).

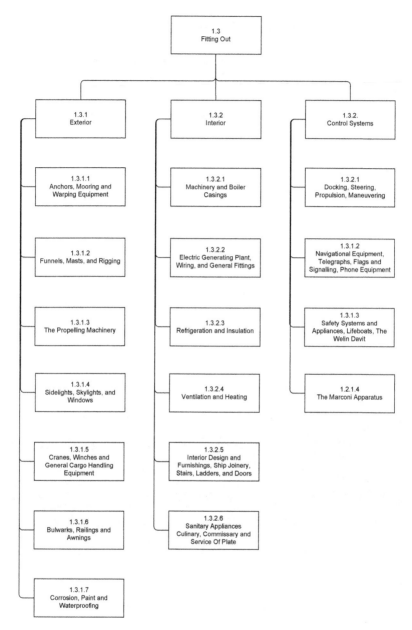

Figure 4.18. Projection of what a possible project work breakdown structure would have looked like for the Fitting Out (level 2). A tremendous number of activities required completion before the ship was ready to sail in a short time frame of less than one year.

Fitting Out the Engine and Boilers and Important Equipment

The most significant work activities included hoisting aboard engines, boilers, and other complex machinery and lowering these into the ship's bowels using Harland and Wolff's huge floating crane (Figure 4.19). Provisions had been made to accommodate this where the nearly completed deck spaces and casings had steel decking and beams that were removable to create openings of sufficient size in some areas. From a **scheduling perspective**, this was one of the most challenging activities.

Figure 4.19. The 200-ton floating crane lifting a boiler on board at the Thompson deep-water outfitting wharf. All 20 boilers had to be lifted into the ship.[20]

Figure 4.20. Fitting engines in the engine room was a complex procedure in very cramped conditions. Note the large group of workers at the back and the engine pillar to the right.[21]

Justifying a Luxurious Interior Finish

The goal was to turn *Olympic* into a floating palatial hotel, where it mirrored opulent "period" designs from Europe. The ship's interior areas were particularly important, especially those very visible to the passengers, especially the first class. The cost of achieving this had been factored into the quality plan, and it was financially justified. However, Harland and Wolff had to deliver a level of quality to what was demanded and not beyond, so as not to require additional project funds.

In today's projects, a cost/benefit analysis determines the pros and cons of quality management activities, calculating the benefits versus the costs. Harland and Wolff would have used a similar approach. Quality has thresholds it has to meet between a minimum and maximum. Overspending on quality, or **gold-plating,** is the over elaboration of functions that incurs additional costs and affects the project schedules.

The project team should strive to deliver what was expected. Under-spending on quality would have delivered under elaborated functions and been rejected.

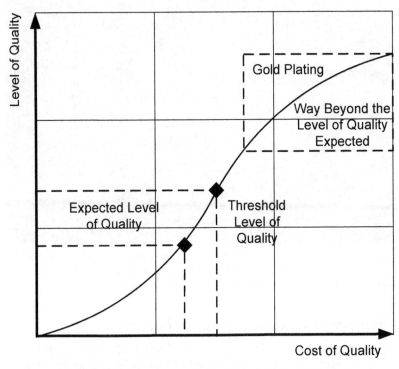

Figure 4.21. Quality management required Harland and Wolff (the supplier) to find the right balance between the desired levels of quality versus the cost (what the project could afford). There comes a point in a project where the client (White Star) does not require more than requested. Gold-plating adds extra features that increase the project costs.

Fitting Out the Interior

Harland and Wolff were renowned for quality and had an army of craftsmen operating in the system of trades. One benefit of this approach was that by completing work to a very high-quality standard increased productivity because shoddy work has to be redone. Quality work completed correctly the first time as expected does not require additional funds to redo the work.

The Harland and Wolff craftsmen (joiners, painters, electricians, plumbers, tilers, and carpet-layers) set to work on the ship's interior. They completed the interior design and furnishings, ship joinery, stairs, ladders, and doors. At the same time, carpenters, cabinetmakers, and artisans finished the last details of the opulent staterooms. Plumbers installed sanitary appliances, lavatory and washbasin fittings, and a few private water closets and baths. Fitters installed the massive galleys that would turn out some 6,000 to 10,000 meals, some 4 courses, daily. Electricians wired the ship and installed more than 10,000 electric lights.

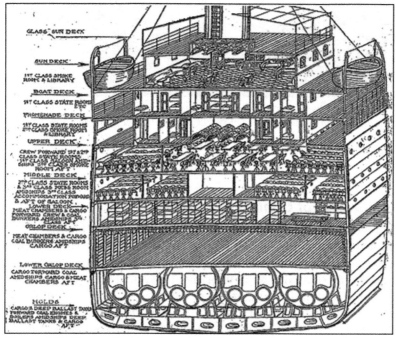

Figure 4.22. Shows the 10 decks, named from the bottom upward: Lower orlop; orlop; lower, middle, upper, saloon shelter; bridge; promenade; and boat deck; or alphabetically named A, B, C, D, E, F, and G. The passenger decks were the promenade deck, bridge deck, shelter deck, saloon deck, tipper deck, middle deck, and lower deck.[22]

Figure 4.23. Olympic's lavish interior and the investments in first-class passenger comfort in the first-class dining room, not just used for dining but also for balls, galas, and entertainment. It could seat 532 passengers simultaneously. The opulent staterooms all followed a period design like the luxury European hotels of the time.[23]

Integration of Control Systems

The Olympic-class ships were sophisticated, modern liners designed to operate an array of semi-automated and automated systems, including systems required in the docking, steering, propulsion, and maneuvering of a ship. There were also systems related to navigation, telegraphs, flags and signals, phone equipment, as well as safety systems and appliances, lifeboats, and the Welin Davit. One of the most important systems was the Marconi Apparatus that enabled communication ship-to-ship and ship-to-shore.

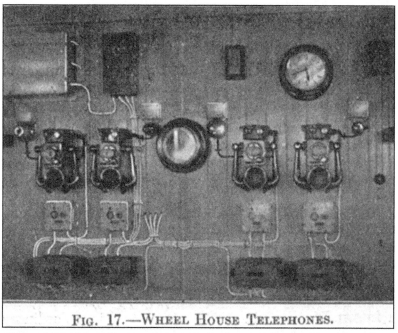

FIG. 17.—WHEEL HOUSE TELEPHONES.

Figure 4.24. Shows the investments in the latest technologies such as the Wheel House telephone used for communication from the bridge to critical lookout points on the ship, such as the bridge wings, crow's nest, and the forecastle (front point on the bow), which could all house extra lookouts.[24]

FIG 18.—ENGINE ROOM TELEPHONES.

Figure 4.25. The phones had very advanced features such as light indicators and extra bells; some phones like this one in the engine room had battery backups. A telephone exchange on board routed calls among 50 phones throughout the ship from the engine room to some first-class staterooms.[25]

Fitting Out the Exterior

There was much work still to complete on the exterior, but because of the limited space under the gantry, it could not be completed during the construction phase. All upper superstructures, such as the bridge, wheel house, funnels, masts, and lifeboats, had to be fitted out.

Figure 4.26. Shows Olympic in dry dock and the fitting of propeller shafts. The Stern frame was a very complex part of the ship, and it had to support the rudder, center shaft boss, and tail end shaft and frame the aperture in which the center propeller revolved. [26]

Figure 4.27. Olympic nearing the end of the fitting out. The single row of lifeboats (20 in all versus the 48) was a compromise, but it was important to Bruce Ismay to give passengers a greater sense of space on the boat deck.[27]

Olympic's Basin and Sea Trials

During the last week of April 1911. *Olympic* was moved from dry dock to the deep-water basin. The week of May 2, 1911, *Olympic* went through her basin trials. On May 29, 1911, *Olympic* went through her sea trials. The trials proved the ship was slightly faster and more maneuverable than expected. Francis Carruthers, the Board of Trade Inspector, issued a seaworthiness certificate.

Figure 4.28 Olympic during her sea trial, on the River Lagan, repainted in her final colors.[28]

End of Titanic's Construction Phase

In 1911, between January and May, *Titanic* was completed and readied for her launch much like *Olympic* in October 1910.

Figure 4.29. Titanic readied for her launch in May 1911.[29]

Figure 4.30. Titanic just before her launch. Note the solitary figure on the stern above the scaffolding around the rudder. Large crowds came out for the launch, which would help promote the ship as another successful delivery from Harland and Wolff shipyards in Belfast.[30]

Titanic's Launch

In May 31, 1911, *Titanic* was launched in Belfast and attracted a much smaller crowd but was timed with *Olympic*'s maiden voyage in June 1911 to maximize media attention. Even though about 100,000 people were present, it is said most were marveling over *Olympic* and paying less attention to *Titanic*. During *Olympic*'s construction, more than 150 photographs were taken, whereas only 45 photographs were taken of *Titanic*'s construction. This lowered interest in *Titanic* was going to be a challenge for Bruce Ismay.

Figure 4.31. An admission pass to Titanic's launch. The launch was a major publicity stunt that would help promote the ship to a public eager to see the output of a four-year project that many had strong connections. The public was very much a project stakeholder.[31]

LAUNCH OF THE TITANIC.

The Titanic the largest ship in the world was launched at Belfast on Wednesday.

Twelve o'clock struck from some distant tower of the city. Lord Pirrie, in a blue suit and a yachting cap, walked up the steps leading to the forefoot of the vast ship, and said a quiet word to the foreman in charge. He came down and vanished along the port side of the ways. Two loud rocket explosions signalled that all was ready. Lord Pirrie pulled a lever, the hydraulic ram gave a stroke or two, there was a slight cracking noise, twenty-five thousand tons of steel parted the timbers under the forefoot and moved slowly away.

" She's gone ! " somebody cried, a cheer was raised, the vast bulk diminished in perspective as the Titanic, without another sound, glided through the forest of steel rods and ties and girders surrounding the ship, and in sixty-two seconds her whole length was in the water, and she was being slowly manoeuvred to her appointed berth.

Thus undemonstratively was born this the most marvellous creature yet conceived by the art of naval architecture and the science of marine engineering .

Figure 4.32. A glowing newspaper account of Titanic's launch highlights the grandeur of the event and reinforces the perception that it was a marvel of modern engineering and architecture.[32]

Publicity and Marketing

The White Star Line publicity booklet of May 1911 published:

> "Today, ships are amongst the greatest civilizing agencies of the age, and the White Star liners *Olympic* and *Titanic*— eloquent testimonies to the progress of mankind, as shown in the conquest of mind over matter—will rank high in the achievements of the 20th century."

This type of communication just further reinforced the claims of invincibility of the Olympic-class ships.

In 1911, the engineering magazine *Shipbuilder* interviewed the project architects who described the ship's principal features, including the emerging technologies used for safety. The magazine later published an article that described the construction project and concluded the ships were practically unsinkable. With a small and niche readership, this article received little outside attention. On June 1, 1911, however, this all changed when the *Irish News* and *Belfast Morning News* contained a report on the launching of *Titanic*'s hull. The description boasted:

> "...the system of watertight compartments and electric watertight doors **and concluded that Titanic was practically unsinkable...**"
>
> —*Shipbuilder* engineering magazine

Bruce Ismay saw this as a further opportunity, which gave him another message in his marketing effort, that of safety, to add to the "largest, and most luxurious, liners in the world." This set up in the public eye a sequence of media events that kept building the perception the Olympic-class ships were unsinkable. To the working class, the ships represented a means to a better life. For the well to do, they represented another symbol of economic power and status.

Figure 4.33. The posters of the Olympic-class ships highlighted how the communications were centered on luxury (excellent accommodation at moderate fares), size (largest steamers), and the implied safety.[33]

Figure 4.34. This IMM advertisement very specifically points to one of the most important safety features available—the wireless telegraph for ship-to-ship and ship-to-shore communication.[34]

Chapter Wrap-up

Conclusion

Although the project construction phase went well, as planned, and hit all the milestones, important compromises from the design phase were not readdressed in the construction phase. The logical explanation for these compromises is that the architects assumed that the "aggregated" safety features remaining would protect the Olympic-class ships from whatever nature handed out. Bruce Ismay perpetuated the situation further by reinforcing (and marketing) with the public the claims of invincibility of the Olympic-class ships, based on these safety features. This further elevated public and passenger expectations about how the Olympic-class ships would perform, and it was timed with *Olympic*'s spectacular launch, which attracted huge crowds, even though the ship was a year from completion.

As the ship's interior took shape during the fitting out, the visible lavish investments in passenger comfort implied that there was an equivalent investment in the less visible ship's safety and operational features. The Olympic-class was marketed to the public as practically unsinkable, which became widely accepted.

Key Lessons for Today

By the end of the construction phase, the project team believed that the safety levels were maintained at the initial levels, and the compromises were forgotten. Therefore, this set a high level of confidence for the project and maiden voyage. Expectations had not been carefully managed with the project team of the changed risk profile.

Today, typically before project managers complete this phase, they would do the following:

- Identify building-block parts versus prefabrication and solution alternatives such as build versus buy.

- Identify nonfunctional alternatives (such as safety features) for the solutions.

- Organize the project to build in cycles. Start with small (prototypes), and then scale up.

- Review government regulations that could affect the project, and plan accordingly. Look to what the regulations should be rather than what they are. Bruce Ismay knew that the lifeboat regulations were going to change in the near term but proceeded with a low number that would have to be up-sized.

- Ensure business executives and sponsors are involved throughout the construction, through steering committees.

- Ensure the business case is reviewed at points through the project and is updated with any changes.

- Examine and carefully assess any emerging "group thinking," which might inhibit asking questions or going outside boundaries.

Educators

Discussion points:

- During this phase, there were already warning signs of problems with the project through the resignation of Alexander Carlisle. Discuss the implications of this.

- Could the change to downsize the number of lifeboats to 16 from 32 be seen as a reduction in scope and advantageous to Harland and Wolff?

- *Olympic*'s successful basin and sea trials further reinforced the project team belief the ships were unsinkable.

- The project communication management strategy continued to perpetuate the invincibility myth of the Olympic-class ships.

- How well was the project phase completed?

- Was Bruce Ismay a (positive) risk seeker? Typically, as the risk increases, the risk seeker's satisfaction increases to where he or she is even willing to pay a penalty to take on projects of high risk.

CHAPTER 5

Testing (Planning and Executing) Phase

In This Time Frame

○ June 14–21, 1911—*Olympic*'s maiden voyage

○ June 1911—*Olympic*'s first incident

○ July 1911—*Titanic*'s maiden voyage is set for March 30, 1912

○ September 1911—*Olympic*'s second incident (*HMS Hawke*)

○ October 1911—*Titanic*'s maiden voyage is delayed to April 12, 1912

○ February 1912—*Olympic*'s third incident

○ March 20–27, 1912—Officers arrive for sea trials

○ April 2, 1912—*Titanic*'s sea trials

○ April 2, 1912—*Titanic* leaves for Southampton

Overview

Olympic went successfully into service in June 1911 but had three serious incidents at sea that had a major impact on

Titanic's project for cost and schedule on the delivery. This highlights the interdependencies that exist with the project and operations period.

Olympic's Track Record

The project team' pride in delivering the largest floating object on the planet was immeasurable. *Olympic* went into service in June 1911 following her sea trials in Belfast. She sailed to Liverpool (port of registry) with the dignitaries aboard from *Titanic's* launch, and she was opened to the public for a day. She then sailed to Southampton and arrived on June 3 to prepare for the maiden voyage. Again, she was opened to the public where the proceeds were handed to local charities such as the Royal Victoria hotel.[1] Many reporters visited the ship to take photographs that were published around the world All this attention proved invaluable in escalating the interest in the Olympic-class ships, and it was part of the project's Communication Management strategy.

> "To mark the advent of the Olympic into the service the pay of the Commodore of the White Star Line has been increased from $5,000 to $6,000 a year, which will be the highest pay in the Atlantic trade."
>
> —Tuesday, 6 June, 1911, *New York Times*

Olympic's Maiden Voyage

On June 14, 1911, *Olympic* finally sailed on her maiden voyage to New York under Captain Smith's command. To highlight the status of *Olympic's* maiden voyage, owner J. P. Morgan was prevented from sailing on her only because he had been invited to attend the coronation of King George V and, afterward, had an invitation to be the guest of the German Kaiser. Because of the marketing, media interest was intense. For example, a full-time reporter from the Times accompanied the crew to record the occasion. *Olympic* arrived in New York with great fanfare. Bruce Ismay was so satisfied with *Olympic's* performance that he could not wait to inform

Pirrie that "*Olympic* was a marvel." On June 21, 1911, Bruce Ismay was encouraged by her public reception, and he told Harland and Wolff that White Star would exercise their option for a third superliner, *Gigantic*. This was even before *Olympic* completed the westbound part of her maiden voyage.

Figure 5.1. Olympic leaves on her maiden voyage, Southampton in the background, on June 14, 1911. Ismay was delighted with Olympic's performance.[2]

Figure 5.2. Photo taken from the aft shows the boat deck top with the first-class promenade deck beneath, with many passengers milling about.[3]

Figure 5.3a. Olympic arriving in New York on June 21, 1911, guided by her tug boats. She was met by great crowds on the dock.[4]

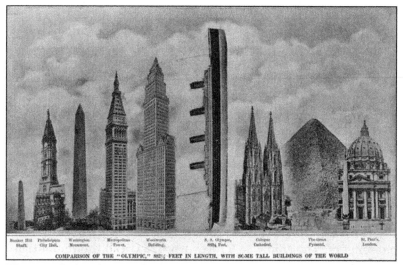

Figure 5.3b. Another article in the renowned Scientific American magazine (July 1911). The article proudly captions the image as "COMPARISON OF THE "OLYMPIC," 882.5 FEET IN LENGTH, WITH SOME TALL BUILDINGS OF THE WORLD."⁵

Quality Management

Thomas Andrews sailed on *Olympic*'s maiden voyage to New York as an architect but to also record notes for possible improvements. This further underlines how important **Quality Management** was to the project and how the philosophy of continuous inspections was carried through into the implementation and operation. Andrews worked very diligently, and no detail or improvement was too small for his attention. This was a good example of the philosophy of continuous improvement, as his notes were applied to *Olympic* and incorporated into *Titanic*.

Olympic's First Incident

Olympic's track record was imperfect, and it was marred by three serious incidents that White Star had to play down. The first occurred on June 28, 1911, as *Olympic* was in New York and was pulled up the North River by 12 tugs and maneuvered into Pier 59 for mooring. The harbor pilot tried to help the

tugs and gave the order for "ahead, dead slow" to nudge
Olympic into the slip. The 200-ton (103-foot) tug *Hallenbeck*
was standing by at the liner's stern when a sudden reverse
of *Olympic*'s starboard propeller drew Hallenbeck in. She
was spun, collided with the bigger ship, and, for a moment,
was trapped under the *Olympic*'s stern. *Olympic* cut off her
stern frame, rudder, propeller, and wheel shaft. The tug's
ensign mast snapped as she was driven into the water and
her decks became awash. The tug nearly sank but was able
to right herself and limp to a pier in the area. The tug owners
sued White Star. White Star, a large organization, was able
to countersue and successfully shift the blame on the small-
time operator of the tug, due to the lack of evidence. On June
28, 1911, *Olympic* left New York with a record number of
passengers. The ship was 92% overbooked, and the round trip
generated a substantial £30,000 ($150,000) profit.

Bruce Ismay Requests Changes

Following *Olympic*'s maiden voyage, Bruce Ismay was very
satisfied with the outcome but realized some improvements
could be made to improve service revenues and positively
affect the business case. Therefore, he suggested two major changes to *Titanic*. Today, as the project management
plan is executed, deliverables are made but also changes are
identified. All projects are affected by changes, and therefore, they require a mechanism for all change requests.
Known as Configuration Management (part of **Integration
Management)**, this protects both the customer from unau-
thorized changes by project staff and the project staff from new
or undocumented requirement changes from the customer.
It controls a product's characteristics. The contract with the
shipbuilder Harland and Wolff had a mechanism that referred
to formal procedures defining how **change requests** were
handled. These changes were late and delayed the project.

Changes to the Functionals

Bruce Ismay's requested changes were functional changes. The
first change was significant, as it replaced part of enclosed "B"

deck promenade with extra first-class staterooms and suites, two with private verandas. This increased the first-class passenger accommodations by 100, a significant increase in revenue of about £12,000 ($60,000) a voyage. It also increased the crew in the catering staff.

The second change included creating the Café Parisian (Figure 5.4), a trellised café overlooking the sea. This highlighted Bruce Ismay's focus on the first class, maximizing the customer experience, and substantially increasing profits for White Star (justifying the changes). It also showed White Star's confidence in the project.

Figure 5.4. Changes required to Titanic included creating the Café Parisian, a trellised café overlooking the sea. This was a significant change and came very late in the project.[6]

From a **risk management** perspective, this is an example of **positive risk** exploitation, in which the positive outcome of the identified risk in increasing sales and customer

satisfaction offset the risk of delay. It highlights that not all
risks are bad and that there are situations where the project
manager should seek positive risks.

Changes to the Nonfunctionals

Thomas Andrews intended to make improvements in the ships
by finding possible flaws in the *Olympic* during her sea trials
and maiden voyage. He had observed a troubling amount of
vibration. Therefore, he began a change to *Titanic* to correct
this, which added reinforcing steel in key areas, including
where the double bottom met the main hull. This was a
difficult change but was very important because it affected
the ship's performance. This is a good example of how change
requests can go beyond just changes to the project scope. For
example, different types of change include preventive action,
corrective action, and defect repairs. In this case, this was a
corrective action, but it could prevent other problems from
occurring.

Titanic Fitted Out

After *Titanic*'s launch in May 31, 1911, there was a period of
close to a year when the ship was fitted out and furnished (the
internal fitting) in the deep-water basin at Harland and Wolff.

By July 1911, planning started for *Titanic*'s maiden voyage
set for March 30, 1912. White Star printed timetables, posters,
and stationary as a build-up to this event.

Figure 5.5. The fitting-out of Titanic in the deep-water basin at Harland and Wolff was the final phase of the project construction, without most of the upper superstructures such as the bridge, funnels, masts, and lifeboats all in place.[7]

Olympic's Second Incident (Hawke)

Olympic's poor track record continued with the second of three serious incidents occurring on September 20, 1911, when she departed Southampton's White Star Dock on her fifth voyage across the Atlantic with 1,313 passengers and 885 crew members. To reach open sea, *Olympic* had to perform the usual reverse S maneuvers off the Isle of Wight. She was under the command of Captain Smith, but on the bridge was Captain George Bowyer, the Southampton pilot to lead the tricky navigation safely through the channels. She was sailing at 14 knots in parallel to the Royal Navy cruiser *HMS Hawke* in a narrow channel 200 yards apart. Both vessels were trying to go down the same stretch of water with not much room.

Commander William Frederick Blunt (Royal Navy) altered the *Hawke*'s position by five degrees to give *Olympic* more room. The 7,500-ton *HMS Hawke* sailing at 16 knots at first overtook *Olympic* but then dropped behind as the liner increased her speed. Suddenly, *Hawke* veered hard to port, and Blunt ordered the port engine stopped and full astern on the starboard. However, it was too late, and *Hawke* rammed *Olympic* head on (Figure 5.6), piercing her outer skin. *Hawke* had a large cement ram under the waterline, which broke off. As *Hawke* finally tore free, she spun around like a top and almost capsized. She had lost balance because of the missing ram and the crushed bow.

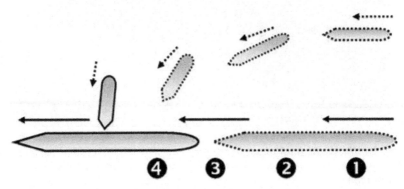

Figure 5.6. HMS Hawke (top right) was drawn by suction forces, known as the Bernoulli Effect, from Olympic (1) forcing the Hawke to turn through 90 degrees (2) and (3) until she hit the Olympic (4) around the bridge area and penetrated the hull further aft.

The damage to *Olympic* was considerable with two areas of penetration. The top area was a gaping triangular hole around 20 feet high, 15 feet across, and 10 feet deep (Figure 5.7). Two of the largest watertight compartments rapidly filled with water, so all the watertight doors were closed. Even with any two compartments open to the sea, the ship would not sink.

Figure 5.7. Two areas of penetration in Olympic's side. The top area was a gaping triangular hole around 20 feet high, 15 feet across, and 10 feet deep. The lower one was right on the waterline.[8]

Figure 5.8. The lower area of penetration in Olympic's side was right on the waterline. Many rivets had popped out in the collision. This was the visible damage, but far more damage was hidden from view.[9]

Initial Inspections

Incredibly, no one was hurt on *Olympic*, as the second-class cabins sliced open were empty because passengers were lunching in the dining room. *Hawke* was equipped with a cement ram that absorbed the impact and diminished her damage but caused extensive damage to *Olympic*. An inspection by the ship's carpenter, chief engineer, and Captain Smith found the ship unseaworthy. The crossing was canceled, and *Olympic* returned to Southampton by way of Portsmouth, helped by tugs. There, her 1,300 passengers were unloaded. Following further inspections in the floating dry dock, the repair work was beyond the abilities of the Harland and Wolff maintenance facility in Southampton. Belfast was the only facility with a large enough dry dock to house *Olympic* to complete the extensive repairs.

Figure 5.9. The HMS Hawke was very much damaged with her bow and front end completely crumpled and her near-front starboard badly buckled.[10]

Today, project managers must look carefully at the track records of previous projects and their implementation success to determine lessons learned and understand some risks. With *Olympic,* the officers and crew were challenged by the ship's large size and handling characteristics.

Patching the Olympic

After the collision, *Olympic* had to be patched and made seaworthy before even attempting the voyage back to Belfast and the Harland and Wolff's repair yard for thorough reparations. She also had to be emptied of cargo, coal, and perishable food stores. From a repair dock in Southampton, it took two weeks to create a gigantic patch made of heavy timbers above the waterline and steel plates below it (Figure 5.10). Many smaller holes were patched up with wood as well. She left Southampton on October 4, 1911.

The patch was placed over the damaged hull plating to seal up the hole, but by the time the ship made it back to the Belfast yard, the patch on her hull had failed and the two aft compartments were again flooded. She arrived in Belfast, 570 miles, at a much-reduced speed of ten knots on October 6, 1911, two weeks after the collision, pulled by tugs and traveling on only one engine.

Figure 5.10. The Olympic after being patched with a wall of timber beams running horizontally. In the center, a twisted vertical frame supports a suspended horizontal pole.[19]

154

Dry Dock Inspection

At Belfast dry dock, *Olympic* was examined, and the damage was very extensive, greater than the first inspection had found. The starboard manganese-bronze propeller blades were badly chipped and unserviceable. There was extensive damage to the propeller shaft, bent out of alignment and inoperable; 18 feet of the outer steel shaft covering was crushed, and the shaft bearings were damaged. The starboard engine's crankshaft was also very much damaged. All this machinery was intended to last the ship's lifetime with little provision for ready replacement. It was also precision milled and took months to procure and manufacture by Harland and Wolff engineering suppliers.

The after-most section was externally accessible, but as the rest was not *Olympic*'s, hull plates had to be removed, not a simple task. She also had a huge hourglass-shaped hole in the plating, and eleven hull plates were damaged above the waterline. Worse still was the hidden damage below the waterline where eight hull plates had to be replaced, including a gash 40 feet wide. Many of the ship's frames were bent and twisted, and thousands of rivets were no longer watertight. *Hawke*'s ram had hit *Olympic* around the bridge area (25% of the ship's length), and it was later recovered from inside *Olympic* (75% of the ship's length).

Figure 5.11. The inspection at Belfast's dry dock highlighted how the Olympic's starboard manganese-bronze propeller blades were very much chipped, damaged, and unserviceable, suffered during her collision with Hawke.[20]

Impact on the Project Business Case

The priority was to get *Olympic* back in service because of the financial pressures to generate revenue. Altogether, she would be out of service for eight weeks, returning November 20, 1911, and missing three transatlantic voyages. This was a huge financial blow for White Star and to the **project's business case**.

Suspending Programmed Activities on the Titanic

From a **Time Management** perspective, all **programmed activities** were suspended on *Titanic*. The workforce had to stop construction on *Titanic*, as she was removed from dry dock to allow repairs on *Olympic*, which took more than four weeks. *Olympic's* plating and channel ribs were replaced, an expensive, labor-intensive, and time-consuming operation.

The *Olympic*'s drive shaft machinery was so very much damaged that the only option was to cannibalize *Titanic* of her machinery, as there was no time to procure this. The Olympic-class ships were not designed for the ready removal of such machinery, so the procedure was difficult, and the risk of causing damage was very high.

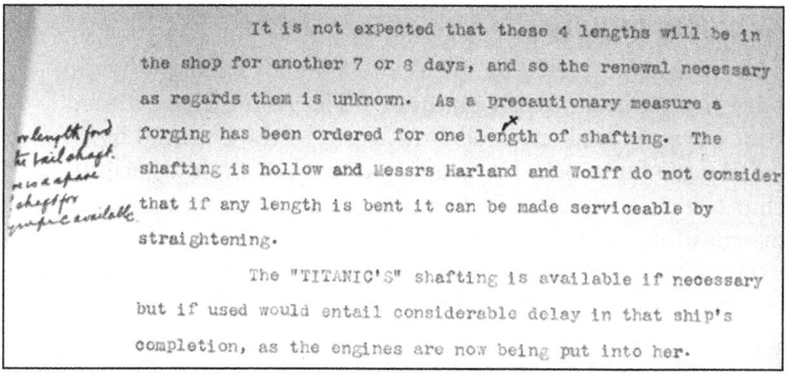

> It is not expected that these 4 lengths will be in the shop for another 7 or 8 days, and so the renewal necessary as regards them is unknown. As a precautionary measure a forging has been ordered for one length of shafting. The shafting is hollow and Messrs Harland and Wolff do not consider that if any length is bent it can be made serviceable by straightening.
>
> The "TITANIC'S" shafting is available if necessary but if used would entail considerable delay in that ship's completion, as the engines are now being put into her.

Figure 5.12. This minute specifies that Titanic's shafting was available as a replacement, but this would considerably delay the project completion. The propeller shaft would have to be removed from Titanic, a difficult operation.[13]

Impact on the Titanic's Schedule

There was a severe impact on the *Titanic*'s **schedule** as workers were transferred to work on *Olympic*. This transfer affected activities on the *Titanic*'s **critical path**, the path with the longest duration in the project, and therefore affected the ability to meet the deadline.

White Star's Credibility Affected

To accommodate the *Olympic*'s repairs, White Star rescheduled *Titanic*'s maiden voyage from March 20 to April 10, 1912, with an announcement in the *London Times*. In addition, the changes Bruce Ismay requested in June required more time. From a **Communication Management** perspective, the negative publicity around the *Hawke* incident was very damaging to White Star's credibility, especially with their principal clientele, the first-class passengers who had to rearrange their calendars for the maiden voyage, something not easy for the world's movers and shakers. This was troublesome for Bruce Ismay whose focus was on making this the most important social event of 1912. He already had the problem of topping *Olympic*'s maiden voyage of June 1911.

Accident Inquiry

The Royal Navy tried to sue White Star for racing *Olympic* past *Hawke* in a narrow channel.[14] The Admiralty claimed that *Olympic* had crowded *Hawke* out of the channel by sailing too far south. International navigation laws stipulate that the overtaking ship must keep clear of the vessel she is overhauling, whereas the latter must keep her course. Naval experts were called, and one sailed from Washington DC, who put the blame on the powerful forces of suction (known as Bernoulli's Principle) exerted by *Olympic,* which was seven times heavier than *Hawke.*

The court passed a verdict for the Admiralty based on the negligent navigation of *Olympic.* White Star, unhappy with the verdict, appealed, claiming *Hawke* had overtaken *Olympic.* Eventually, the inquiry found that the collision was only because of the faulty navigation of *Olympic*'s pilot Captain Bowyer, but the case against *Olympic* was dismissed without cost because of the rule about compulsory pilotage. The Admiralty fought the court's findings through to 1914, taking it to the highest court in the land. White Star had to fight this doggedly to keep its reputation in tact.

Impact on Project Budget

Olympic's repair costs were an astronomical 17% of the original build cost. In total terms, the loss was £250,000, which included lost revenue from three trips. From a **Cost Management** perspective, this was not forecast and outside the project budget. White Star had no option but to face the situation, absorb the costs, and get *Olympic* quickly back into service. The ship was only insured for 66% of the building cost but that didn't matter, as insurance would not cover this incident. Overall, the owners with the two months of lay-up were left with serious financial problems. A level of risk was also associated with rushing *Olympic* back into production. A rush job could introduce errors that could require more time out of service.

Impact on Project Schedule and Cost

The project situation is best shown by the S-curve below (Figure 5.13) with a graphic representation of the accumulated budgeted costs over time. Typically, costs are high up front in such a project and accelerate during execution but taper as the project closes. The Budget at Completion (BAC) is the estimated cost of the project when completed, and it is considered the project baseline. The Estimate at Completion (EAC) is the amount the total project is expected to cost on completion and at this time. So right after the second incident (with *Hawke*), the EAC for *Titanic* went up, and this is known as the Estimate to Complete (ETC), or the estimated additional costs to complete the project. For this, the project would have turned to the **contingency and management reserves** established, which were not part of the **cost performance baseline**, accounting for uncertainty and risks and for any unplanned changes. The Variance at Completion (VAC) is the difference between the total amount the project was supposed to cost (BAC) and the amount the project is now expected to cost (EAC) with VAC = BAC − EAC. In today's projects, this is part of Earned Value Management (EVM).

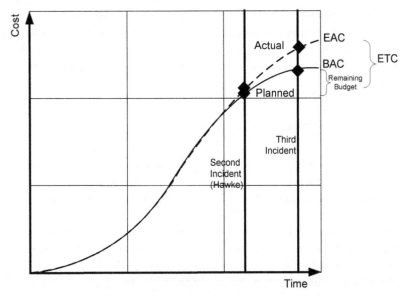

Figure 5.13. An S-curve showing the impact of the two incidents to Olympic on the cost side of the project. The single line represents the planned costs with Budget at Completion (BAC), £1.5 million ~$7.5 million a ship. The dotted line represents the actual costs with Estimate at Completion (EAC), the increase after the two costly incidents. The Hawke incident increased the budget by 17%.

Schedule Crashes

From a **Time Management** perspective, Harland and Wolff had to crash the project to meet the schedule and, specifically, the rescheduled maiden voyage for *Titanic*. There was no additional time through **contingency reserves or buffers**. **Crashing** is a strategy to compress the project schedule without reducing project scope. Crashing uses alternative strategies, such as outside resources, for completing project activities for the least extra cost. In the minutes of 1911, Harland and Wolff's board discussed the hiring of extra rivet-squads.[15] In this case, cost had to give, as the scope could not be reduced; it had already increased with the changes Bruce Ismay requested. Crashing has a potential adverse side effect in that it creates critical paths (the longest duration path). Harland and Wolff had to know this. An alternative strategy is **fast tracking**, in which sequential project activities are overlapped or performed in parallel.

Today, the impact on the project completion date is determined by the **total float**, which is measured by subtracting early dates from late dates of critical path completion.

Resource Leveling

In today's projects, teams use this schedule network analysis technique, once the critical path is determined, to meet both delivery dates and to have a constant level of resource availability and use. This technique reviews the project schedule for over and under allocation of resources; for example, individual resources might be assigned to more than one task. Typically, a project manager uses a **resource histogram**, with all resources used across a period, to eliminate spikes. Another important technique is to create a **resource breakdown structure**, which is very similar to a work breakdown structure (see Figure 5.14). It decomposes project resources by category, phases, or type to give a visual mapping of resource types required, by logical groups.

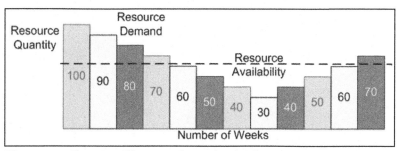

Figure 5.14. Resource Histogram is used to level out over and under allocation of resources.

161

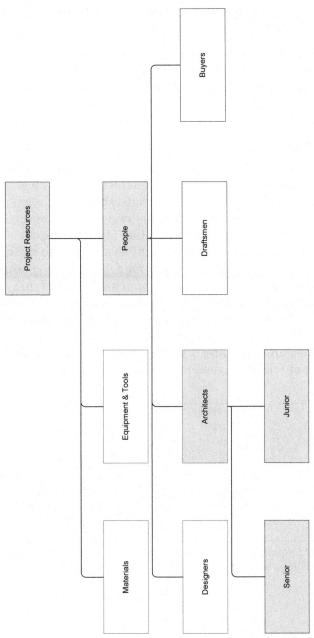

Figure 5.15: A resource breakdown structure is very similar to a work breakdown structure. It provides a visual map of resource types required, by logical groups.

Examining Previous Similar Projects

The *Hawke* incident was significant because not only did it severely affect the project, but also no insurance was paid, and White Star could not recover the repair cost from the British Navy, as the court ordered that each party pay its damages. The *Hawke* incident haunted Captain Smith, however; he continued to claim his innocence by saying that *Hawke* had struck *Olympic,* and the collision was the fault of *Hawke*'s captain. He had a significant non-collision bonus to protect.

Project managers today should closely examine how previous similar projects have delivered solutions into service, taking particular note of any lessons learned, or even conducting post-mortems. Any failures or anomalies, procedures used, and organizations involved need to be closely examined. This should be input to the overall planning.

Olympic's Third Incident

Olympic's poor track record continued with the third of the three serious incidents.

Loss of Propeller Blade

This occurred on February 24, 1912, again under Captain Smith's command. She lost a propeller blade on her port side during the eastbound crossing, the return leg of the round trip. It was likely knocked off by a well-known wreck in the Grand Banks floating beneath the surface about 750 miles off Newfoundland. The unbalanced propeller was disengaged. *Olympic* sailed with two propellers to Belfast for repairs, and *Titanic* again was switched out of dry dock on February 29, 1912, to allow a replacement blade to be fitted.

LINER *OLYMPIC* HITS A WRECK
Chicago Daily News, Tuesday 27 February 1912

Ship Due at Southampton on Its Way To Belfast for Repairs.

Belfast, Ireland, Feb. 27—The White Star liner *Olympic,* which left New York on Wednesday and was due in Southampton today, struck a submerged wreck in the Atlantic early this morning and is on her way to this city for repairs.

According to latest reports, the damage to the ship was confined to the propeller. She will land her passengers at the usual points before proceeding here.

The *Olympic* carried a large passenger list and many notables were included among her first CAbin passengers. Among them were Ambassador Reid, the Duke of Newcastle, Count Apponyi, and W. E. Corey. [16]

More Problems

The work was completed on March 2, 1912. *Olympic* was then hauled out of dry dock and turned 180 degrees where her port bow was grounded. She had to be put back into dry dock for examination, wasting more time. She was back in service March 7, 1912, but had to cancel the scheduled transatlantic trip of March 6, 1912. The fitting of the replacement blade should have taken two days, not the six days it took. Harland and Wolff had one set of replacement blades. If these were for *Titanic*, which had a slightly larger blade, this would have required the fitting of all three blades because the blades all had to be the same size, otherwise, the arrangement would be out of balance.

Figure 5.16: Olympic in dry dock after the refit of a new propeller blade after the third incident. The workers stand proudly in the background. When she was hauled out port, her bow was grounded, and she had to be put back in for examination, wasting more valuable time.[17]

Crashing the Critical Path

Harland and Wolff desperately tried to complete the project on schedule to deliver *Titanic* on April 1, 1912. The **critical path** was further crashed as additional workers were hired boosting the workforce from 14,000 in May 1911 to a peak of 17,275 (10,000 at work in the yard on Queen's Island, Belfast, working three shifts) just a month before the deadline. During March, even before *Titanic* was completed, many engineers who would be operating her arrived to familiarize themselves with the systems and machinery.

Increasing the workforce had many concerns such as the space available for workers to operate in safely and productively. It required great skill and much experience, as there was a point in the project where the "output-to-effort ratio" became saturated. Adding workers became counterproductive. This is the law of diminishing returns, and Harland and Wolff knew this.

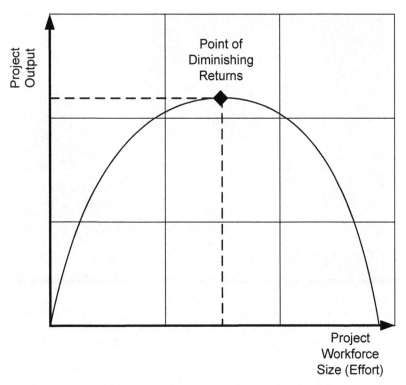

Figure 5.17. Harland and Wolff knew the law of diminishing returns whereby increasing the workforce reaches a limit, and the "output-to-effort ratio" becomes saturated.

Bruce Ismay Requests a Further Change

Just four weeks before *Titanic*'s maiden voyage, Bruce Ismay pushed for the final change, which was major and unusual for a ship almost ready for service. The change was to glaze over the forward third of the open promenade deck "A" with sliding windows, as first-class passengers complained of spray from the wind and sea. This change was so late that it is difficult to comprehend the bravado with which it was requested. It is also difficult to comprehend why Harland and Wolff would accept this change, and it indicates a very likely breakdown (compromise) in **change control** and **verification scope**.

Making the Delivery

From a scheduling perspective, Harland and Wolff delivered *Titanic*, remarkably, a day late, but the question is at what cost. It did not affect Harland and Wolff, as according to the *Irish Times*:

> "This was a boom period for Belfast shipbuilding, with regular increases in tonnage output. Between 1910 and 1912, Harland and Wolff launched 23 ships in addition to *Olympic* and *Titanic*. By the end of the 1910–11 financial year, the company's profits totalled nearly £110,000 (equivalent to about €7.3 million in today's money), and by the close of 1911, the number of men at work in the shipyard had risen sharply to almost 15,000. The company's weekly wage bill was about £25,000 (€1.6 million today), a sizeable figure for the time. The shipyard's tremendously impressive output would have been impossible without efficient practices, a skilled and well-organized workforce, forward planning, and the international business expertise of the Harland and Wolff chairman, Lord Pirrie."[18]

Harland and Wolff's delivery of *Titanic* only a day late is a testament to Harland and Wolff's project management skills and the importance of **Human Resources management** to the project's success, namely, the ability to expand the work-force and maintain productivity. By all accounts, Harland and Wolff had created a project environment conducive to work in, in which there was much pride for people in being associated with the project.

Titanic Sea Trials

As *Titanic*'s fitting out neared completion, Harland and Wolff planned sea trials (tests) before she was passed to White Star.

Meeting Contract Conditions

Both organizations needed to be assured that the project deliverable would meet the conditions and requirements laid out in the contract. Testing gave the shipbuilder an opportunity to make any necessary adjustments and avoid the risk of financial penalties or having the ship sent back to the shipyards.

Formal Acceptance of Deliverables

From an **Integration Management** perspective, the **verify scope process** gets formal acceptance of completed project deliverables from stakeholders, typically the project sponsor. It determines whether the deliverables conform to requirements and documents the formal acceptance and sign-off of these. Unlike the **perform quality control process** (part of **Quality Management**), which validated the deliverables, it focuses on the acceptance criteria of the deliverables or the scope of work. In today's projects, the deliverables should also tie nonfunctional requirements such as performance that could link to service-level objectives established earlier in the project. The business case determines the solution costs and the service level required based on how much risk the organization can tolerate. Testing should assess how well the solution will meet these service-level objectives and identify any gaps.

Acquiring an Operations Leadership Team

White Star needed an operations team to plan and execute the sea trials. Today, when testing, we typically **acquire an operations team (Human Resource Management)**, which is then developed and managed. Likewise, White Star created an operations team that included the captain and the senior officers. They were employed full time, and they had vastly better pay; for example, the captain was paid £1,250 pa, plus a £200 pa non-collision bonus, and the officers were paid £300–500 pa. Between March 20 and 27, 1912, all the officers (chief, first, and juniors) arrived for the sea trials. The team was critical in planning and executing the sea trials and accepting the final deliverables.

Planning Sea Trials

This typically consists of outlining a plan for several types of testing. For example, *Titanic* had to be operationally tested for seaworthiness, checked for stability, and carefully assessed for weight and loading details. One test was the "incline test" that checked the ship's weight and center of gravity using a simple inclining experiment. It also checked earlier manual calculations. Other tests included the dock-side trials held for the preliminary testing of main and auxiliary machinery. Formal speed trials were normally necessary to fulfill contract terms. These required achieving a specific speed under specific conditions of draft and deadweight.

Ensure Testing Is Broad

Likewise, projects today need to outline plans for testing the solution, and this needs to be done for both the functional and the nonfunctional requirements. However, the focus should be on testing the incredibly important nonfunctional requirements because they define a system's operational characteristics. The testing needs to be dynamic, building on earlier static testing (or walkthroughs). It is important not just to test at unit level but also the overall solution, partly because of the perceived high costs of simulating a complete environment. Therefore, only partial testing is ever completed, which lowers confidence for a successful implementation. It is critical not to implement the solution with the view that any problems will be flushed out by the user or in the short term. This risky approach might put the business in serious jeopardy.

Determine Alternatives for Implementation

Projects today need to outline plans for alternatives should the implementation run into problems. For example, with *Titanic*, two alternatives were postponing the trip or completing the crossing with known faults.

Compromising on the Sea Trials

White Star deemed *Olympic's* track record adequate to launch an almost identical sister ship straight into service without extensive sea trials. The extensive first-of-class trials were carried out on *Olympic* that defined the class 'standard'. As second-of-class all the shipowners really needed to know was whether *Titanic* was markedly different to Olympic.

Olympic's track record reinforced the shipowner's perception that *Titanic* was ready for her maiden voyage. However, this only involved comparing the physical structures of the two ships and did not look at the readiness of the crew (people) or procedures (processes).

Assessment of Business and Technical Risks

Many projects today can make a similar mistake of building a false sense of security by not examining similar implementations closely enough. They need to complete an assessment of the business risk and technical risk of an impending launch. An assessment determines which tests are required, why the tests should be undertaken, and the test objectives and overall testing strategy.

The extensive tests performed should include stress, performance, regression, security, and operational testing, which requires planning test cases for each testing objective, that is, what is tested, how it will be tested, with what, and what are the expected outcomes or results. Most important, these test cases should test the solution's nonfunctional features. Planning should determine how the tests are to be performed, in what environment, and who should objectively perform the tests. For example, the builders should not test their own work; rather, independent test teams should do so.

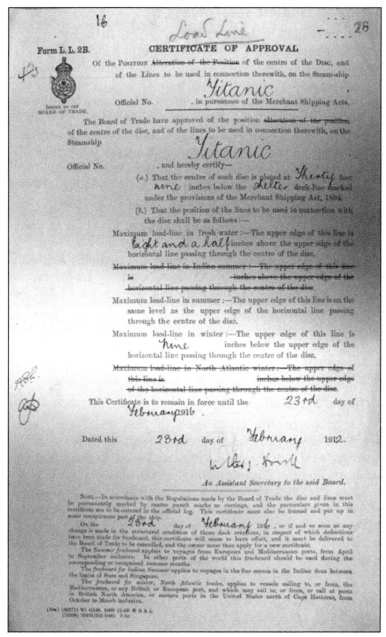

Figure 5.18. Outline of the BBOT certificate of approval for the position of the load lines of the ship in different water and conditions, which was one of several certificates passed on to Harland and Wolff.[19]

Executing Limited Sea Trials

The sea trials had been planned for April 1, 1912, and some 79 of her "scratch" crew traveled from Liverpool and Southampton. However, with bitter, cold weather and strong winds, the team **assessed that the risk** of damaging the hull was too great. With improvement in the weather, the sea trials started on April 2, 1912, under the auspices of Captain Smith. Not only the project team, but also the public at this point, perceived that Olympic-class ships were invincible. This, coupled with the shipowner's perception of *Olympic's* solid track record, ignoring the three incidents, enforced the thinking that *Titanic* was ready for the maiden voyage.

Titanic underwent less than one day of sea trials to meet the deadline for the maiden voyage. During the sea trials, the project team was eager to compare the new ship's speed to *Olympic*. *Titanic* averaged 18 knots for a two-hour run, running up to 21 knots. *Titanic's* turning circle was 3,850 feet with a forward motion of 2,100 feet. Emergency stop from 20 knots took 850 yards (unloaded). This was in contrast to the *Olympic,* which underwent more far-reaching sea trials that included engine tests and adjustments to her compasses.

Certificate to Sail

As part of the **quality assurance,** the Board of Trade surveyor Francis Carruthers (external inspector) performed a **quality audit** to determine how well some government standards and regulations, primarily the seaworthiness of the ship, were met. Carruthers was satisfied, and he issued the safety certificate and declared her "Good for one year." In his notes, during his final inspection of *Titanic,* he indicated hydraulic riveting (considered superior to riveting by hand) in three-fifths length of the shell plating.

Outcome of Sea Trials

But what was missing from *Titanic's* sea trials? Not only were the trials not extensive, but they were also not considered as critical as they were with the *Olympic*. Basin trails were never completed, where as *Olympic* went through four weeks of these. As a result, *Titanic* was not put through incline tests or any complex handling maneuvers such as "S-turns" used to get around hazards in an emergency. The officers also had very little time to acclimatize to the ship's handling. Officer Murdoch complained about the time required to get used to *Titanic*.

Figure 5.19. Titanic passed her very limited sea trials with little trouble and got the seaworthy certificate from the British Board of Trade. The pressure was to stay on schedule and put her into operation.[20]

Was the Contract Fulfilled?

The core project team and stakeholders were convinced that *Titanic* was to go into operation. *Olympic* was used as a test bed or yardstick for *Titanic*, and much faith was put in her track record. However, it is debatable how well the experiences learned from *Olympic* were applied to *Titanic*, especially when her track record was viewed so positively. In addition, the sea trials were used to determine how well the builder Harland and Wolff had delivered the project.

Normally, the four-week basin and sea trials gave the builder an opportunity to make adjustments and to deliver to specifications. From a **Procurement Management** perspective, these trials' successful completion allowed both sides to verify that the contract's conditions were satisfied. White Star likely gave leeway on this because of the excellent relationship and trust with Harland and Wolff. Even though handover papers were signed by Harold Sanderson (White Star) and Thomas Andrews (Harland and Wolff), there has to be a question over whether the contract was truly completed and what oral agreements were made.

Guarantee Group

Headed by architect Thomas Andrews, a group of eight handpicked workers (best men in their field) was responsible to see to any unfinished work or use their skills collectively to solve any problems or document these for completion when the ship returned to Belfast. Andrews had traveled on three other ships' maiden voyages: the *Adriatic*, *Oceanic*, and *Olympic*. This was particularly important with *Titanic* because of the shortened trials.

Decision to Implement

It is important in today's projects for project managers to protect the testing phase's integrity, so it is extensive and ensure it covers the nonfunctional requirements defining the solution's operational characteristics. For the more technical reader, this determines things such as the run-time availability, security, and system management, as well as non-runtime scalability, portability, maintainability, environmental factors, and ability to evolve. The nonfunctional requirements ensure the solution delivers the functions for which it was designed.

Implement Only with a Green Light

The lesson from this for projects today is to move into an implementation schedule only according to preagreed plans, so the organization (and more specifically, the operations team) is well prepared and ready. It is important to assess any new business changes through a **change control process** (part of **Integration Management**) that carefully examines the risks, as in an earlier launch date.

Planning to Sail

The discussion on going ahead with the maiden voyage was probably short. The business pressures for *Titanic* to sail were enormous. This was understandable, considering the large investments tied up in its four-year construction. In addition, *Olympic* had been out of service for several months in dry dock for repair because of the collision with *HMS Hawke* (Figure 5.6), so substantial revenue had to be made up.

The decision to sail was made on the same day as the sea trials, and *Titanic* weighed anchor at 8 p.m., and a mad dash to Southampton followed. There was no opening of the ship to the public in Belfast.

Chapter Wrap-up

Conclusion

White Star's overall business objective was to have two luxury liners crossing paths on the Atlantic on a weekly schedule. As *Olympic* had been twice unexpectedly out of service, for more than 10 weeks, Bruce Ismay was eager to see *Titanic* move into service as quickly as possible to generate new revenue. White Star played down the seriousness of *Olympic's* incidents. Harland and Wolff went to great lengths to complete *Olympic's* complex repairs and at the risk of not completing *Titanic's* contract on time. *Titanic's* schedule was put in jeopardy, so the project workforce had to be rapidly expanded. It is a testament to Harland and Wolff's skill in project management that they could pull this off. It also signifies how important **Human Resources management** was to the project's success.

From a **Cost Management** perspective, the *Olympic/Hawke* collision was deeply unsettling, with high repair costs. No insurance was paid out, and White Star could not recover the repair cost from the British Navy, as the court ordered that each party pay its damages. The third incident highlighted that the design should have considered various repair situations with access to replacement equipment and the need for spare parts.

With *Olympic* in service for 11 months, White Star deemed *Titanic's* maiden voyage as low risk. After all, the two ships were the same class and nearly identical. There was great confidence in *Titanic* and the maiden voyage. However, no two ships have identical handling characteristics. As *Titanic's* project moved to the next phase, little had changed in the project team's or public perception that the ship was practically unsinkable, as White Star carefully managed expectations.

Key Lessons for Today

In today's projects, it is important to plan adequately for testing and prevent any compromises to the solution as it is implemented into production, for example, bowing to business pressures to get the solution into production quickly. White Star was very much driven by the pressing economical need to move *Titanic* into service. In reality, *Titanic*'s testing consisted of the maiden voyage across the Atlantic while loaded with passengers. The decision was understandable, considering the huge amounts of capital that had been invested and remained tied up during construction. Today, typically before project managers complete this phase, they would do the following:

- Review similar previous projects with the Project Management Office (PMO) for common problems.

- Plan the level of testing; select the right tests and acceptance criteria.

- Assign an operations team the ownership and control of the testing. Ensure they are brought early into the project, are involved in planning and executing the testing, and are given incentive to focus on what is important, such as service-level agreements, especially when third parties or outsourcers are involved.

- Define alternatives to launch (withdrawal) and back-out plans if the implementation is unsuccessful.

- Create a test environment that mirrors the live environment.

- Use computer models to reduce test time, complexity, and cost.

- Always deploy in a test environment first and run it parallel to live environment until a degree of confidence is established.

- Ensure testing is broad, not just on functional requirements, but also on nonfunctional requirements.

- Avoid any "in-house" implementation processes (change management) that seem incomplete and not well supported by the overall project (or organizational) governance.

- Run risk assessments before all project phase gates.

- Ensure test teams are independent and unbiased with incentives to test objectively.

- Ensure that first design assumptions are fully tested.

- Establish the ability for an implementation to be stopped if any testing criteria fail.

- Ensure that major testing can be halted if required.

Educators

Discussion points:

- Did Bruce Ismay's changes strain the relationship with Harland and Wolff, in particular, the last set of changes in February (to glaze over the forward third of the open promenade deck "A")? Did Bruce Ismay understand the change's cost, the implications the change would have on the project schedule, or the overall risks to the project?

- Did White Star gamble in not adequately planning and executing the testing? What was the potential cost?

- Why did Captain Smith believe in the invincibility of Olympic-class ships even after the three incidents?

- Would a reasonable alternative to the maiden voyage have been sending *Titanic* across the Atlantic with just a skeleton crew?

- At what point is a project completed? How did this apply to this project?

CHAPTER 6

Implementation and Operations Phase

In This Time Frame

o April 2, 1912—*Titanic* rushed from Belfast to make second tide into Southampton.

o April 3, 1912—*Titanic* arrives in Southampton in preparation for maiden voyage.

o April 4, 1912—*Titanic* docks at White Star berth #44; some crew are added.

o April 5, 1912—*Titanic* dressed in flags for Good Friday at Southampton.

o April 6, 1912—General cargo and rest of crew are added.

o April 8, 1912—Fresh food arrives, final preparations overseen by Thomas Andrews.

o April 10, 1912—*Titanic* departs Southampton at 12:00 en route to Cherbourg.

o April 10, 1912—*Titanic* arrives at 6:00 p.m. at Cherbourg.

o April 11, 1912—*Titanic* arrives at 6:30 a.m., Queenstown, Ireland, next stop New York.

Overview

This chapter looks at how White Star went forward with the maiden voyage. Today, this critical phase requires following a well-planned, well-thought-out, and well-prepared implementation process, with diligence, checkpoints, and assessments as the solution is implemented into production.

Proceeding with the Maiden Voyage

White Star was well aware that *Titanic*'s sea trials were rushed, with failure to complete meaningful testing, but it got them the all-important Board of Trade certificate to sail. They were willing to take on the risks introduced by the limited sea trials based on the overall confidence. Typically, **Risk Management** is performed throughout the project (Figure 2.18) through the **monitor and control risk process**. The identification of risks is an iterative process that occurs with the project team's help. White Star probably assessed that the *Olympic* track record and experience could be adequately transferred to *Titanic* by sailing her under the same captain and officers. The Guarantee Group would address any defects missed in the sea trials.

Commander of the Fleet

As a show of confidence, White Star promoted Captain Smith to the Admiral of the White Star Fleet, as he took command of the flagship of the fleet. This was a great honor for him and enhanced his reputation. Ship's captains were important in bringing back passengers who had a previous memorable voyage on their ship. From a **Human Resources management** perspective, this promotion underpinned his leadership and the company's faith in him.

Managing Risks

One known risk in the Atlantic was Iceberg Alley, a path of ice that breaks off the coast of Greenland, and it is carried south into the Atlantic. In the notorious Iceberg Alley, ice flows on well-established currents in often-dense formations.

It is tens of miles wide and up to one hundred miles long. White Star captain and officers knew well that the volume of ice had increased in 1912, with warmer weather patterns likely due to the El Niño of 1911. The probability of the risk was high, so officers decided to move the sailing path south by ten miles, enough to avoid the most concentrated areas of the alley, but not significantly enough to affect the sailing time. This is a good example of **qualitative risk analysis** in **risk management,** which involves assessing the probability and impact of the risk (consequences to the project), determining the timing and urgency, and prioritizing the risk. Typically, there are four **risk response** strategies:

- **Accept** the consequences of the risk and develop a contingency plan.

- **Avoid** by eliminating the cause, so the risk no longer exists.

- **Transfer** the direct risk to a third party, an insurer, clients, or vendor.

- **Mitigate** reduces the probability of the risk occurring and its impact of the risk.

In this situation, White Star could not avoid the risk completely, so they mitigated it.

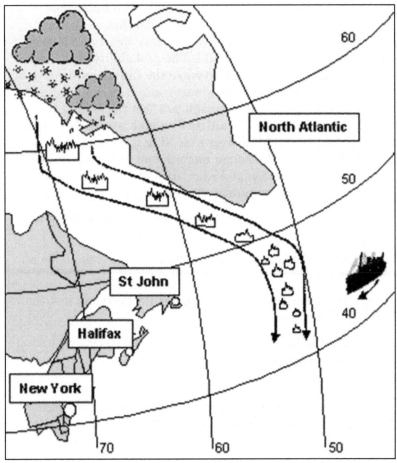

Figure 6.1. Iceberg Alley, a path of ice that breaks off the coast of Greenland, and it is carried south into the Atlantic. Because of the El Niño of 1911, White Star decided to move the sailing path south by 10 miles.

Acquiring an Operations Crew

The White Star crew was hired just before a sailing. They were well paid but only when the ship sailed, so this was a challenge to many a crew who had to find odd jobs. This approach did not bode well for **developing and managing a team**, typical when **acquiring an operations team (Human Resources Management)**, as they had little time together before the voyage started. There was much work for a crew to do to prepare, especially in familiarization of the new systems on the ship. Thomas Andrews briefed the department heads, showed crew members how to operate various things, and completed troubleshooting any problems that occurred.

A crew of 900 was required to run the ship and consisted of hospitality workers (around 475 stewards, bellboys, maids, housekeepers), engineers (around 320 mechanics, firemen (stokers), trimmers, greasers, coal porters), and mariners (83). The former were paid per sailing £3 15 and relied on tips. The latter two were paid £5. The living conditions were a challenge with 40 sleeping to a dormitory. The crew fell under an organizational hierarchical structure. For example, per watch, 7 engineers supervised the work of more than 90 firemen, trimmers, and greasers. In addition, there were always 3 deck officers on duty per watch.

Preparation to Sail

There were only five days of preparation time between April 4th and 9th. This period was compressed, as much had to be done, including the loading on board of fuel, provisions, water, and medical stores. The situation was compounded by a coal strike that, if it had continued just a few days longer, would likely have canceled the scheduled sailing on April 10, 1912. White Star was desperate to keep to the schedule, so coal was sequestrated from other ships—*Oceanic, Adriatic,* and American liner SS *New York*—with the subsequent transfer of passengers. There was much pressure to turn around the new ship. *Titanic* was never opened to the public like the *Olympic* because there was no time. This was a lost opportunity to promote the ship further from a **communication management** perspective.

Figure 6.2. The coal shortage forced passengers to be transferred from the White Star liners Oceanic, Adriatic, and SS New York to Titanic, as coal was sequestrated from these liners for Titanic's maiden voyage.1

Executives on Board

To ensure a smooth transfer of ownership, Bruce Ismay decided he would be on board for the maiden voyage, along with Harland and Wolff's chief architect Thomas Andrews, as with *Olympic's* maiden voyage in June 1911. J. P. Morgan would travel as well. This was to have a dramatic impact later in the story.

Marketing Titanic's Maiden Voyage

Olympic's maiden voyage of June 11, 1911, had attracted worldwide media attention, and more than 100 articles were published. Before *Titanic's* maiden voyage, media attention had waned, and only 30 articles were published. This must have concerned Bruce Ismay, as a successful maiden voyage helped guarantee good continuing ticket sales. With the staggered construction delivery of three ships, Bruce Ismay realized he had a marketing opportunity where he could make the case that each ship was an improvement over the last. This is a typical expectation of a technology-oriented society. By beating *Olympic's* best crossing time, he could market *Titanic* as a superior liner.

Figure 6.3. Titanic in Southampton proudly decked out in flags on Good Friday (April 5, 1912). Pressures on meeting the deadline prevented the ship from being opened to the public. This was a missed marketing opportunity with the public and press.[2]

Setting a New Service Level

To promote this fact, before *Titanic*'s maiden voyage, Bruce Ismay put a very small shipping announcement that would appear in the *New York Times* in the edition to be printed on Monday, April 15th. It stated that *Titanic* would arrive on Tuesday night rather than Wednesday morning, White Star's published schedule.[3] This was a publicity stunt. He was effectively creating a new Service-Level Objective (SLO) for *Titanic*. Alarmingly, Bruce Ismay did this single-handedly without verifying it with his captain and officers. This and his forthcoming actions proved fateful in pushing the ship to its operational limits.

Who Should Set Service Levels?

The lesson from this for projects today is that SLOs should not normally come under the control of the business, instead the operations team with the mandate to run the services in production. The business certainly has a role in helping define what the SLOs should look like, but this should be carefully negotiated and agreed.

Improvements in Ships

In the United Kingdom, the Olympic-class symbolized a peak in British shipbuilding and engineering against the growing competition. In truth, there were no improvements to *Titanic*'s power output and speed over *Olympic*. Both were identical, apart from some cosmetic differences with some functions. Bruce Ismay would race *Titanic* across the Atlantic at maximum speed and beat *Olympic*'s best time, set over 11 months of passage.

This New York Times headline and article from April 10, 1912, builds the tension that Titanic will be arriving in New York. It underlines the widespread perception that Bruce Ismay had built with Titanic; she was the largest vessel afloat and an improvement over Olympic.

THE TITANIC SAILS TO-DAY.
Largest Vessel In World to Bring
Many Well-Known Persons Here.
Special Cable to The New York Times.

London, April 9—The White Star liner Titanic, the largest vessel in the world, will sail at noon tomorrow from Southhampton on her maiden voyage to New York.

Although essentially similar in design and construction to her sister ship, the Olympic, the Titanic is an improvement of the Olympic in many respects. Capt. Smith has been promoted from the Olympic to take her across. There are two pursers, H. W. McElroy and R. L. Baker.4

Figure 6.4. Manufacturers, who sought this opportunity, saw product placement and association with the project very positively.[5]

Transferring Knowledge and Experience from Olympic to Titanic

White Star was very aware of transferring knowledge and experience between ships. What is significant is that Captain Smith, First Officer Murdoch, and Second Officer Lightoller were part of the ship's crew on *Titanic's* maiden voyage, as they were with *Olympic*. It is also a further example of **risk management** where risks were mitigated through having the right level of experience transferred. Today, this would be part of the training plan (**Human Resources Management**).

Titanic Narrowly Avoids Serious Collision in Southampton

On April 10, 1912, *Titanic* had a very near collision on leaving Southampton on departure for her maiden voyage. *Titanic* left to great fanfare with large crowds. As *Titanic* sailed through the harbor, she had a very near collision with the steamer *SS New York*. *Titanic* was sailing at speed, and as she passed *SS New York*, the three-inch steel mooring lines could not stand the strain and snapped. The ship broke her mooring, swung out, and was within four feet of colliding with *Titanic* (Figure 6.5). Fortunately, Captain Smith put one of the three propellers into reverse, and one tugboat (*Neptune*) managed to secure a line to the steamer. Together, these actions prevented a serious collision.

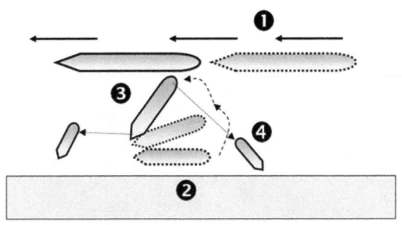

*Figure 6.5. Titanic and the steamer SS New York nearly collided. Titanic
(1) dragged SS New York (2) off her mooring, breaking retaining ropes.
SS New York (3) came within four feet of colliding with Titanic. The tug
Neptune (4) prevented a major collision by restraining SS New York.*

In many ways, this was very similar to *Olympic*'s collision
with *HMS Hawke,* and this illustrated the challenges the
officers and crew faced in operating such a large ship in
channels and near other ships. It also highlighted the pressure
the officers were under in navigating the ship so hastily
through the port. Had the collision taken place, the maiden
voyage would have been canceled. We can only speculate about
the extent of the potential damage.

Figure 6.6. This photo shows Titanic sailing past SS New York. The motion dragged SS New York (left of center) stern first off her mooring, next to Oceanic (left of her). The tugboat (far left and front) is Neptune that managed to secure a line to SS New York.[6]

Figure 6.7. A photograph of Titanic and SS New York within an ace of colliding (photo and caption taken from the Daily Mirror), shows how close she got, within four feet of a collision. Neptune (tug in the center) pulled SS New York away (left of center) from Titanic (right of center). The event showed the intense media coverage of the maiden voyage.[7]

Pick Up at Cherbourg

Titanic arrived at Cherbourg to pick up passengers en route from Paris. Most important, J. P. Morgan did not board as planned because of alleged illness. The more likely explanation was he had installed a young French mistress in a villa he had taken at Aix-les-Bains. Morgan's planned presence had helped encourage the sale of first-class tickets, and there were 53 millionaires aboard. The most prominent passengers of the time included millionaire John Jacob Astor IV and wife Madeleine Force Astor; Macy's owner Isidor Straus and wife Ida; industrialist Benjamin Guggenheim; Denver millionairess Margaret "Molly" Brown; Sir Cosmo Duff Gordon and wife couturiere Lucy; George Dunton Widener, his wife Eleanor, and son Harry; journalist William Thomas Stead; cricketer and businessman John Borland Thayer, wife Marian, and 17 year old son Jack; the Countess of Rothes; author and socialite Helen Churchill Candee; United States presidential aide Archibald Butt; author Jacques Futrelle and wife May; silent film actress Dorothy Gibson; and producers Henry and Rene Harris.

Still Testing

The abruptly short sea trials skipped critical tests such as the incline, navigation equipment, complex handling maneuvers, and getting around hazards in an emergency. The short time line prevented officers from properly acclimatizing to the ship's handling characteristics, which meant that they had to catch up en route.

One of the most important lessons for today is that documentation created at each phase of the project is then timely transferred as knowledge to the operations team.

Inspections at Queenstown

On April 11, 1912, *Titanic* reached Queenstown, Ireland, the last port before the main Atlantic crossing, and the Board of Trade inspectors boarded *Titanic*. As part of the **quality assurance**, they performed a final inspection (**quality audit**) to determine the vessel's seaworthiness, checking the hull, boilers, and machinery. They also checked for provisions, water, fuel, and medical stores. They looked over the steerage compartments for light and air and inspected the health of the crew and steerage passengers.

Fudged Lifeboat Drills

An important part of the inspection was safety. A lifeboat drill was performed in front of the Board of Trade inspectors to determine the crew's operational readiness. During the drill, only two of the twenty lifeboats were lowered, but they did not reach the water, so the test was not completed. The drill outlined that it took eight to ten well-trained men to prepare and lower a lifeboat. More important, the test failed to highlight how poorly the crew was prepared to handle a disaster requiring the launch of all twenty lifeboats. There were only eighty-three mariners in a crew of nine hundred. The rest operated either a luxury hotel or the ship's machinery, and they were not sailors. After the inspection, *Titanic* received a report of "seaworthy and ready to sail" from the Queenstown Board of Trade, so officially, they were good to sail.

The lesson from this for projects today is that the purpose of testing is to bring to light major flaws missed in the requirements gathered in the planning, design, and construction phases. The final test (lifeboat) in Queenstown was fudged, which highlighted the whole approach was still influenced by the belief that little could go wrong.

Ensuring a Support Structure Is in Place

For a project today, implementing a solution into production, it is essential, that a support structure is in place for monitoring, operating, and controlling the solution through a combination of tools, procedures, and people. The goal is to meet the Service-Level Objectives (SLOs) set out in earlier project phases to guide the architects in the solution design. Typically, the support structure is multi-hierarchical (four levels) and built around a "recovery clock," (Figure 9.2), where the speed of recovery is essential to getting the service back to normal. It is important that a support structure is factored into the project and is designed, created, and tested through the respective project phases.

Titanic's Support Structure

A ship's support structure was not much different from a four-level support structure for today's solutions (see Figure 6.8). *Titanic* had a frontline that included the lookouts in the crow's nest, the on-duty officers, and the crew on the bridge. The radio operators in the wireless room, analogous to a help desk or call center, communicated with the outside world, controlling the flow of information to and from the ship. Secondary support staff included the safety officer (plotting positions of icebergs, currents, and weather systems), the navigation officer (plotting the position of the ship and maintaining the course), and specialized technical positions (the carpenter and ship's doctor). The first and second officers and the captain provided tertiary support. Likewise, today's projects need to ensure a similar (up to three-tier) support structure is in place for the provided service. Such a structure is typically made of operators, help desk, call center, technical and specialist support, operations leaders, and managers.

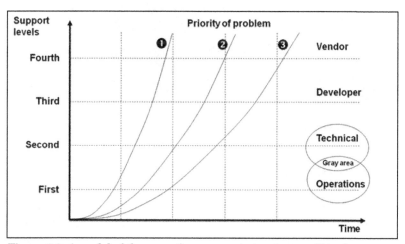

Figure 6.8. A model of the operations support structure on Titanic was similar to a four-level operations support structure today. Incoming problems are prioritized and dealt with escalating levels of support.

Leaving Queenstown and Assessing the Risks

As *Titanic* left Queenstown, Ireland, on April 11, the captain and officers were fully aware of the dangers ahead in the Atlantic, namely, storm and icebergs. The winter of 1911/12 had been exceptionally mild, and many icebergs had broken loose. Typically, **Risk Management** is performed throughout the project through the **monitor and control risk process**. White Star knew that the volume of ice spotted had increased in 1912, likely due to the El Niño of 1911. Therefore, they took preventive actions and decided to move the sailing path south by ten miles, enough to avoid the most concentrated areas of the alley, but not significant enough to affect the sailing time.

Figure 6.9. Third-class passengers thronged the stern of the ship as she left Queenstown, Ireland, waving to relatives and friends at quayside.[8]

Strategy for the Crossing

Bruce Ismay would race *Titanic* across the North Atlantic. He was confident in not only the ship and its safety systems, but also in the ship's operation and the early warning systems in place. *Titanic* had built-in feedback mechanisms for warning the officers and crew of changes in the environment. These included visual monitoring from the crow's nest and bridge, the ice detection test, and ice warnings from other ships in the vicinity. As far as Bruce Ismay was concerned, these feedback mechanisms would provide enough warning to take evasive actions.

Risk identification is an iterative process that occurs with the project team's help. White Star probably assessed that the track record could be adequately transferred from *Olympic* through the captain and officers. The Guarantee Group would address any defects.

Chapter Wrap-up

Conclusion

The ship went into service with the maiden voyage in a hectic and rushed manner after only a week of preparation, preceded by a very rushed "testing" phase. This was the second of two great sister ships, and a perception had already emerged about the superiority of the Olympic-class ships. Even if things went wrong operationally, the ship had enough safety features to protect it in any scenario. This instilled a mindset in the officers, crew, and passengers that the ship was unsinkable. Why else were there 53 millionaires aboard? In many ways, Bruce Ismay's marketing effort was too successful.

On board, some operational risks related to the voyage were managed, such as moving the sailing path through Iceberg Alley further south or transferring the same officers and captain from the *Olympic*. Some risks were introduced, such as Bruce Ismay's tinkering with service levels, trying to beat *Olympic's* best crossing time. He certainly should not have had the mandate to do this without the consent of the operations team. This was a turning point, as this was where Bruce Ismay usurped the control of the ship from Captain Smith. It was also reflected in *Titanic's* reckless navigation through Southampton.

Key Lessons for Today

By the end of project, the view evolved that the Olympic-class ships were huge lifeboats. Such was the confidence in the ship's safety that by this point in the project, disaster recovery and business continuity plans were considered superfluous. In short, the people "who should have gotten it"—the naval architects—allowed the compromises to pass.

In today's projects, a project manager preparing for this phase would work closely with the operations team to do the following:

- Undertake business and technical risk assessments before implementation.

- Avoid giving developers rights to a live production environment.

- Ensure an adequate support structure (four levels) is in place for the solution delivered and the service provided.

- Ensure the boundaries are clear between the project team and the operations team over the deliverables and implementation.

Educators

Discussion points:

- Should Bruce Ismay have been on board the ship for the maiden voyage because of his position as chairman of the board (Captain Smith reported to him)? If so, in what capacity?

- Bruce Ismay's presence usurped the reporting hierarchy of Captain Smith and officers working for him.

  o Could Bruce Ismay have been prevented from compromising *Titanic's* operational mandate?

  o Could Bruce Ismay have been dissuaded or prevented from sailing on the maiden voyage?

- What leadership styles (Autocratic and Directing, Consultative Autocratic and Persuading, Consensus and Participating, Shareholder and Delegating) did Bruce Ismay and Captain Smith each exhibit?

- What types of power did Bruce Ismay and Captain Smith each use (Formal, Reward, Coercive, Referent, Expert)?

- Speculate on the relationship between Bruce Ismay and Captain Smith.

CHAPTER 7

Warning Signs Ignored

In This Time Frame

○ April 11, 1912—*Niagara* strikes ice and news reaches *Titanic*.

○ April 11–15, 1912—*Titanic* gathers speed and races across the Atlantic.

○ April 14, 1912, 7:30 p.m.—Three ice warnings are received and passed to the bridge indicating ice field is 50 miles away (captain is entertaining guests at dinner).

○ April 14, 1912—9:30 p.m.—Lookouts instructed to be careful of growlers.

Overview

This chapter looks at how the maiden voyage progressed. On the surface, from a passenger's perspective, the ship's operation ran like a well-oiled machine, but under this façade, operational problems brewed.

Common Implementation Mistakes

Titanic was now fully on her way. The maiden voyage differed from the other routine voyages, as the world's press and media were paying close attention to its success. This put additional pressure on the operations team, and in this atmosphere, the risk of making mistakes increased.

Niagara Runs into Ice

Through the wireless Marconigram, the captain and officers were made aware of the fate of the French liner *Niagara* that had run into ice in Iceberg Alley (Figure 7.1) on April 11th. According to a Reuters report, *Niagara* sailed through thick fog at reduced speed, and there came a "severe shock" of striking ice.

> ## FRENCH LINER IN THE ICEFIELD.
>
> ### PASSENGERS' ALARM.
>
> NEW YORK, April 16.
>
> The French liner Niagara, on arriving here today, reported that on Wednesday night she was approximately in the vicinity where the Titanic sank. She ran into a field of ice, and was so badly bumped that she sent out the wireless distress call, "S.O.N." Thick mist prevailed at the time. The ship was running at reduced speed, and had been brushing against small icefloes for some time, when there came a severe shock. Those sitting at dinner at the time were thrown from their seats to the floor, dishes and glassware were scattered over the saloons, and the stewards were thrown down. The scared passengers rushed on to the decks in swarms. The captain made an inspection, and subsequently sent out a second wireless message, saying that he could proceed to New York under his own power. —Reuter.

Figure 7.1. Coverage of the Niagara incident (on April 11) by Reuters. Titanic's captain and officers were well aware of the incident shortly after it was first reported.[1]

The following description of *Niagara's* incident was published:

> "Passengers were hurled headlong from their chairs and broken dishes and glass were scattered throughout the dining saloons. The next instant there was a panic among the passengers and they raced screaming and shouting to the decks..."I thought we were doomed," said Captain Juham yesterday. "At first I feared we had been in collision with another vessel as I hurried to the bridge. But when I saw it was an iceberg and that we were surrounded by ice as far as we could see through the fog, my fears for the safety of the passengers and the vessel grew....I am sure Captain Smith had a similar experience in practically the same locality when the *Titanic* went down."
>
> —*New York Herald*, April 17, 1912

Niagara was holed twice below waterline and had some of her plates buckled. The captain inspected his ship and found that, although it was leaking, due to the buckling of plates, it was in no immediate danger, where upon he sent a wireless saying he could proceed to New York under his power. He subsequently determined that he could reverse the propellers, pull the ship off the ice, and make his way backward. The ship on arrival showed little damage, although some water was in the hold.

Reaction on Titanic to Niagara's Collision

On board *Titanic*, the *Niagara* incident probably didn't make much impact on the captain and officers; collisions like this were very common in the Atlantic. However, Captain Smith handed the *Niagara* telegram to Bruce Ismay, probably to warn him of the dangers ahead and possibly to hint at slowing down. Bruce Ismay read it to himself and then put it in his pocket, which was noted in front of witnesses.

Mid-Atlantic Turn

More and more boilers were lit as the voyage progressed and the machinery was 'run-in.' With the heavy reported ice during the voyage, Captain Smith delayed the mid-Atlantic turn to take a more southerly course by 30 minutes. With crossing times, the paying passengers and the public expected captains to press on and meet the published schedule, very much like airlines today. This is a further example of how **risks were managed**. A safer course of action would have been to delay for 60 minutes, but this would have affected the crossing time. Captain Smith knew from firsthand experience how captains had been forced to move the southern track 60 miles further south in 1903, 1904, and 1905 because of ice conditions.

The lesson from this for projects today is the importance of identifying and assessing new risks as they occur through the project, and typically, this **(Risk Management)** is performed through the **monitor and control risk process**. Risk identification is iterative and occurs with the project team's help.

Beating Olympic's Best Crossing Time

Passengers later said that, during the voyage, they heard Director Bruce Ismay pressuring Captain Smith to go faster, to arrive in New York early. The primary reason was to generate some free press about the new liner. One first-class passenger, Mrs. Emily Ryerson, was returning to New York for her son' funeral. She overheard Bruce Ismay and Captain Smith discuss the crossing time and beating *Olympic*'s best crossing time. She revealed that she and her friend, Marian Thayer, had met Bruce Ismay on deck on the fatal day and been shown by him an ice warning.

Concerned over getting to New York one day early and finding a hotel room at night, she asked Bruce Ismay, "Will you be slowing down?" Bruce Ismay replied, "Certainly not, we will put on more boilers to get out of it." Bruce Ismay's concern, Mrs. Ryerson said, was with crossing the Atlantic in record time.[2] Passengers also claimed to have seen Bruce Ismay flouting an iceberg warning at dinner time, waving it around, and then placing it back in his pocket. Bruce Ismay later denied the story at the US inquiry, but Fireman Frederick Barrett disclosed that three additional boilers had been lit, more than at any other time in the journey, and the ship reached its peak speed. Bruce Ismay risked the ship's safety by forcing it at top speed into the treacherous Iceberg Alley.

THE ILL-FATED S. S. TITANIC, FOUNDERED APRIL 15, 1912.

Figure 7.2. *Titanic's speed was steadily increased across the North Atlantic, as her machinery was 'run in', to beat Olympic's best crossing time.*[3]

Operational Feedback Mechanisms

The Olympic-class ships came with very important feedback mechanisms to alert the officers and crew when the ship was near ice and ice fields.

Feedback Mechanism—Wireless Technology

Titanic came with the latest in wireless technology known as the rotary spark discharger, only developed in 1912. It gave *Titanic* a much stronger signal than other ships. Over the next two days, *Titanic* received eight warnings reporting ice, icebergs, and ice floes. The radio operators, however, only sporadically relayed these ice warnings to the bridge because they were preoccupied with the flood of outgoing commercial radio messages and congratulatory incoming messages. Marconi employed the radio operators, and they did not fall directly under the auspices of the ship's hierarchy.

As Marconi company employees, they were contracted primarily to relay personal messages and were paid for each commercial radio message sent. Effectively, this was a toy for the wealthy first-class passengers sending messages as a status symbol of their wealth (fee to send a wireless telegram was twelve shillings and sixpence/$3.12 (or $62 today) for the first ten words and nine pence per word thereafter). This inundated the radio operators who were run off their feet. The radio operators overloaded by commercial traffic (noise) did not timely pass the ice warnings (signal) along. In the 36 hours between leaving Southampton and the collision, the *Titanic* received and sent 250 passenger telegrams. Unfortunately, there were no procedures to prioritize messages, so a telegram to friends in New York had greater priority over any warning messages.

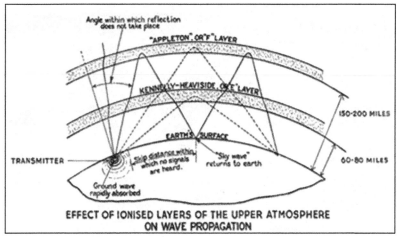

Figure 7.3. The Marconigram or wireless communication was an important development for the safety of ships as it allowed ship-to-ship or ship-to-shore messages. This magazine article explained how it worked.[4]

The lesson from this for projects today is to plan that the operations team leverages built-in feedback mechanisms to warn of impending problems. These mechanisms collect data, and they are programmed to sense "events" or "warnings" that alert operations staff or trigger rules-based software. Any external warnings, potentially from customers or suppliers, need to be taken seriously and investigated. Finding the meaningful information in a "sea of noise," or redundant information, is invaluable. With *Titanic*, if someone had pieced together all the ice-warning information, it would have indicated a giant ice field around 80 miles wide, directly ahead. Effectively, there was no accurate macro-view of the environment surrounding the ship.

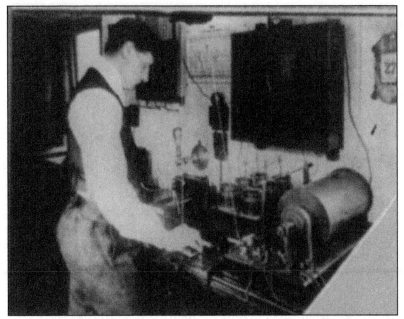

Figure 7.4. The radio operators were Marconi company employees contracted primarily to relay personal messages. They were paid for each commercial radio message sent.[5]

On April 14, five ice warnings were received during the day. The last one that Captain Smith saw was at midday, a second from *Baltic* indicating "Large icebergs and field ice 42 N 51 53' W." At 7:30 p.m., three ice warnings were received and delivered to the bridge, but Captain Smith was below at dinner. The ice field was now 50 miles ahead as the ship passed through ice-strewn waters at full speed. At this speed, *Titanic* would indeed arrive in New York on Tuesday night.

Feedback Mechanism—Crow's Nest

Titanic had some built-in visual monitoring through the crow's nest and the bridge. Beyond the two lookouts in the crow's nest, Officer Lightoller maintained a lookout himself from the bridge. *Titanic* carried six specialist lookouts with two per four-hour shift, and the next shift change was to start at midnight. A question remains why no extra lookouts were on duty, given all the warning signs. A minimum number of lookouts were posted. It was typical to post extra lookouts on the ship's bow, a great vantage for ice, to which a telephone link ran from the bridge (see Figure 7.5).

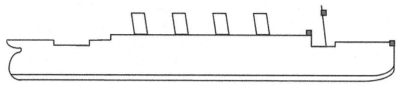

Figure 7.5. Aside from the crow's nest and bridge, there was a lookout point on the ship's bow where extra lookouts could have been posted. It would have been very effective in that night's environmental conditions. This is another example of the overconfidence of Captain Smith and officers.

Figure 7.6. Titanic's crow's nest was about 95 feet above the water. It was connected by phone with the bridge. This was the only system in place that could still prevent a collision.[6]

Change in Conditions

The lesson for today's projects is that external factors can affect the ability to deliver services. For example, there are critical periods in the business cycle, such as month-end processing, where conditions change and get very busy, increasing the likelihood of something going wrong and leading to failure. These periods require more diligence where "extra lookouts" should be posted. Unusually quiet conditions should not be taken lightly, and they are more likely a sign that something has gone wrong.

Missing Binoculars

At the voyage's outset, *Titanic*'s lookouts discovered that their binoculars were missing, which was very unusual. It was customary always to have at least one pair in the crow's nest. The lookouts had repeatedly reported this to the officers since leaving Southampton. The officers did nothing about this, even though they could have shared their own. The lookouts were resentful at not having binoculars because they were the tools of the trade required to make them effective. Explanations varied from the statement that they were assigned to officers on the bridge to the speculation that someone stowed them away and could not find them in such a large ship. The truth we now know was Second Officer David Blair was taken off the *Titanic* at the 11th hour, and he forgot to leave a key to open a locker containing the pair of binoculars for the designated lookouts.[7]

The lesson for today's projects is that certain staff in critical positions, such as operations, should have the best equipment available, over others, despite where they are in the organizational hierarchy.

Figure 7.7. Newspaper report (post-disaster) highlights the missing binoculars story, which came out in the inquiry and shocked the public.[8]

Feedback Mechanism—Ice Detection Test

A steep decline in seawater temperature is a very accurate guide to the proximity of ice. Normally, when approaching ice fields, tests were taken by drawing seawater from over the ship's side with a canvas bucket, and then placing a thermometer in the bucket. When repeated, these tests accurately indicated the proximity of large ice floes. However, one passenger noticed a mariner filling the bucket with tap water. When asked why, he explained the rope was not long enough to reach the sea. The ice detection test was fudged probably because he might have thought he would have been reprimanded for the problem.

The lesson for today's projects is to investigate feedback mechanisms reporting data contrary to other feedback data collected. In *Titanic's* story, the mechanism was faulty, but rather than report the problem, the data was falsified to cover it up. It also emphasizes the importance of testing all feedback mechanisms and operational procedures before implementation.

Sea Conditions on the Night

Any experienced mariner would recognize sea conditions indicative of ice fields. The sea is calmer, as the bergs, ice floes, and pack ice dampen water movement. The seawater also takes on an oily appearance as it approaches freezing point. On the night of April 14th near Iceberg Alley, stars brightly illuminated the sky, and the sea was incredibly calm. The cold weather created a haze on the horizon. These conditions made it difficult to outline the horizon, as it merged with the sky. The officers must have perceived that anything would be seen well in time with such "excellent" visibility. The overall visible distance that objects could be seen from the ship was thought to be beyond the norm, giving the officers a high level of confidence in being able to spot all hazards in a reasonable time.

Chapter Wrap-up

Conclusion

The ship's speed was steadily increased across the Atlantic. *Titanic*'s officers had little time to familiarize themselves with the ship, and they were not as well prepared as they should have been. As *Titanic* approached the ice field, operational risks were taken because the operational feedback mechanisms were not working properly, and they had been compromised. The mariner responsible for the ice detection test readily admitted a problem to a passenger, but did not report it to the officers. The officers didn't share their binoculars with the lookouts probably because of their elevated position in the hierarchy. All this indicates that a level of distrust existed between crew and officers.

Today, many projects severely compromise implementation by not taking it seriously enough and bowing to business pressures to get the solution into production quickly. Bruce Ismay was very much driven, not just by the pressing economical need to move *Titanic* into service, but also to get an attention-grabbing result by beating *Olympic*'s best crossing time.

Key Lessons for Today

In today's projects, a project manager preparing for this phase would work closely with operations to do the following:

- Ensure changes are *not* introduced "blindly" into production.

- Ensure change process has executive support and "teeth" to be effective.

- Ensure change process strategies exist for rapid implementation of important fixes.

- Help refine service-level objectives and agreements once the solution is in production.

Educators

Discussion points:

- Why didn't the officers share their binoculars with the lookouts? Was this an arrogance based on having formal naval training, that it was more important they carried the binoculars (status symbol) than the lookouts?

- Were *Titanic*'s sea trials (testing) extended into the maiden voyage? Was this acceptable, considering the huge amounts of capital that had been invested in the project?

- With *Olympic*'s lost revenues, Bruce Ismay was eager to see *Titanic m*ove into service as quickly as possible. The business *opport*unity was to have two luxury liners crossing paths on t*he Atla*ntic on a weekly schedule. But was this worth the risk?

- Could more have been done to spot problems with the feedback mechanisms?

- Speculate on the relationship between officers and crew based on the ice detection test problems and the mariner's inability to report it to the officers.

- How well was the project phase completed?

CHAPTER 8

Failure in Operations

In This Time Frame

○ April 14, 1912, 11:05 p.m.—*Californian*, after a near-fatal collision, sends an ice warning to *Titanic*.

○ April 14, 1912, 11:33 p.m.—Lookout Fredrick Fleet reports a "dark mass" on horizon to the bridge.

○ April 14, 1912, 11:40 p.m.—The lookouts notify the bridge "Iceberg right ahead!"

Overview

This chapter looks at *Titanic*'s final hour before the collision. The feedback mechanisms had gone wrong, but there was ample warning for the captain and officers to act. Because no further preventive action was taken, the collision became almost inevitable.

Contact with Californian

At 11:05 on the night of April 14, the ship *Californian* was north of *Titanic*, bound for Boston. After a near-fatal collision with an ice shelf, Captain Stanley Lord, clearly rattled by the near miss decided against proceeding and pulled up for the night. The risk-averse captain chose to pull up for the night, as his primary responsibilities were to his ship. He was also following the order of his shipping line:

> "Commanders must run no risk which might by any possibility result in accident to their ships. It is to be hoped that they will ever bear in mind that the safety of the lives and property entrusted to their care is the ruling principle that should govern them in the navigation of their vessels, and that no supposed gain in expediting or saving of time on the voyage is to be purchased at the risk of accident."
>
> —Written orders from *The Leyland Line*

Message Not Received

Surrounded by pack ice, but in no danger, *Californian*'s radio operator Cyril Evans, under direct orders from Captain Lord, sent an ice warning to *Titanic*. Most important, he did not start it with the conventional Captain *Californian* to Captain *Titanic* to convey the warning's importance, but with a more casual:

> "Say, old man, we are stopped and surrounded by ice."

Titanic's radio operator Jack Phillips had been working a 14-hour day sending/receiving commercial traffic. He also had been up in the early hours repairing the transmitter with Harold Bride. He failed to recognize the importance of the telegram, as it was preceded by the casual "old man," instead of the more formal captain-to-captain sign. Phillips responded with the infamous:

> "Shut up, shut up, I am busy. I am working
> Cape Race, and you are jamming me."

Evans, frustrated, did not try again, turned off his wireless, and went to bed at 11:30 p.m. The message Evans was going to send contained particular information about the density of the pack ice (bergy bits, growlers, and full-sized bergs). The field ice stretched for three miles, and it was almost impassable. Lord did not go to bed, but stretched on the settee, all dressed, as he expected to be called to the bridge.

Figure 8.1. Californian's risk-averse Captain Lord chose to pull up for the night of April 14 after a close collision with ice was averted. His primary responsibilities were to his ship.[1]

This last ice warning received from the Californian was not passed back by Phillips to the bridge because it was incomplete and Phillips was overloaded with work. The procedure for passing messages back to the bridge was confusing at best. Radio ice warnings were not timely passed to the bridge, if at all like the last ice warning received, from Mesaba.

Captain Smith's Failure to Act

On the night, the air temperature went down to 33° F, or 1° C, and was noticeably cold for the passengers. First Officer Murdoch was concerned about the cold's impact on the fresh-water tank. Captain Smith did not respond to all the warning signs of cold. He did not post additional lookouts on the forecastle or the two bridge wings. He also failed to warn engineers to be ready for emergency orders from the bridge.

FIG. 15.—FORECASTLE TELEPHONE. FIG 16.—POOP INSTRUMENT.

Figure 8.2. The phones provided rapid communication between the bridge and forecastle, crow's nest, and poop. Flag and signal lamps glowed for an incoming call in case noise levels were high.[2]

Captain Smith was somewhat resistant to technology and relied on "gut" feel and experience. He undermined the significance of Marconigram information and wanted a second source to verify it before he took any action.

Failure in Feedback Mechanisms

Titanic's early warning system had failed because of the failure to report problems with key feedback mechanisms. With the ice detection test, this was possibly because the mariner feared being blamed for the problem, although with any new system, there are teething problems. This, coupled with general overconfidence in the ship's safety, apathy to the fate of the French liner *Niagara*, and inaccurate information on the extent of the giant ice field, led to a state of gross indifference. No one had expected icebergs to be directly in the ship's path at this point in the voyage, as icebergs did not usually drift down as far south as *Titanic*'s course. Finally,

Bruce Ismay's pressure and demand to beat *Olympic*'s best crossing time (new SLO), pushed *Titanic* to her highest speed and past her operational limits.

Heading for a Collision

Titanic was heading for a collision. It was almost inevitable. The ship, close to its maximum speed, raced through icy still waters littered with small bergs and pieces of ice. The lookouts, without binoculars and a freezing wind hitting their eyes, tried to outline the horizon through the haze common in these conditions. As they struggled to make out the shape of a dark mass looming in front of them, they delayed reporting this to the bridge.

Preparing Operations for Support

The lesson for today's projects is that in monitoring a new solution just implemented into production, the operations team needs to be very familiar with the solution and the service. In a project **Human Resources Management** requires a staffing management plan to develop the training to prepare an operations team for the implementation (the maiden voyage) and post-project operational support. The plan should ensure that the operations team could pro-actively prevent failures from happening to the solution and ensure the service meets its service levels. They need good visibility into the solution and the environment. They need to quickly assess and analyze data in front of them, collected from feedback mechanisms set up during the project's planned testing phase. As the mechanisms become noisy, they need to diagnose situations and determine deviations from set norms, any potential impacts, and overall extent. They need to clarify whether something is actually wrong or just problematic. They need to make the right decision whether to escalate and at what priority.

Ice Spotted by Lookouts

At 11:30 p.m., *Titanic's* lookouts determined a dark mass. After much deliberation, they determined that it was a "growler," or "black iceberg"—an iceberg that has flipped over and is dark in color. With a calm sea and no breakers against the base of the growler, it was nearly invisible in the haze.

At 11:33 p.m., lookout Fredrick Fleet reported a "dark mass" on the horizon to the bridge by phone. First Officer Murdoch took the call, acknowledged the sighting, and requested the lookouts to provide regular reports of what they saw. Thinking the dark mass was a passage line through the ice, Murdoch did not alert the captain. The dark mass was likely a gap in the field ice ahead. The lookouts delayed raising an alarm. Only when they were close enough, did the haze turn out to be densely packed ice, and the dark mass an iceberg. This had now turned into a critical situation.

At 11:40, once sure of the sighting, the lookouts notified the bridge with the infamous "Iceberg right ahead!"

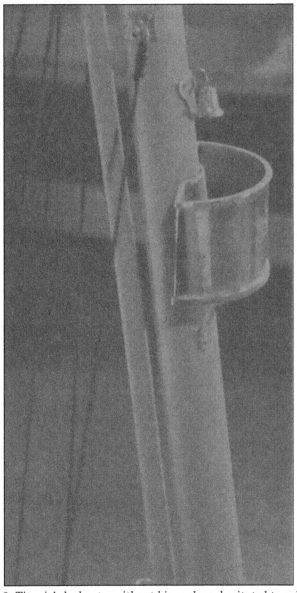

Figure 8.3. Titanic's lookouts, without binoculars, hesitated to raise an alarm, as they struggled to see through the haze.[3]

Murdoch Tries to Avoid a Collision

First Officer Murdoch, chief duty officer, calmly took the call
and with his binoculars confirmed the sighting about 900
yards ahead. From all the evidence available today, Murdoch
took the following actions:

First, he cut power to the engines. This made sense, as
putting the engines into reverse would just churn up the water,
affect the rudder, and limit the ship's steering and handling
capability. Besides, only two propellers could be reversed.

Second, there was not enough distance to stop the ship, and
he could not get around the iceberg. Therefore, he attempted
a port-around or an S-turn, first steering hard a-port, and then
hard a-starboard trying to decelerate the ship sharply (Figure
8.4). With only 40 seconds of reaction time (2.5 to 3 times
the ship's length), this would bring him parallel to the iceberg,
rather than a head-on collision.

Third, he threw the electric switch to close bulkhead watertight
doors (Figure 8.5) as a precaution. In hindsight, these were
probably the best possible actions he could TAke.

As *Titanic* swung back to starboard, Murdoch came very close
but just failed to clear the ice, and he and the bridge staff
braced themselves for a collision.

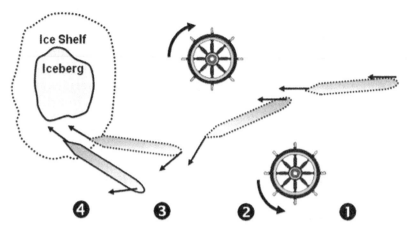

Figure 8.4. Murdoch's action was to port around the iceberg. At (1), Murdoch throws the wheel hard to starboard, moving the ship left. At (2), the bow slowly reacts and moves left, while the stern continues forward. At (3), Murdoch throws the wheel hard to port, moving the wheel right. At (4), the bow slowly reacts and moves right, while the stern continues moving left, away from the iceberg. The bow runs aground on the underwater ice shelf, and the ship grinds to a halt.

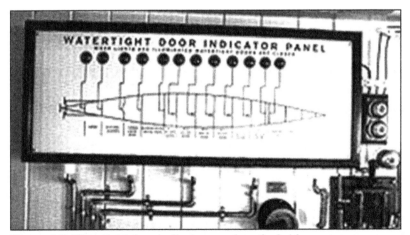

Figure 8.5. Watertight door indicator panel on the bridge was an important safety feature that showed the status of the doors, and it was used in an emergency to close the watertight doors, seal off compartments, and prevent flooding across multiple compartments.[1]

Reacting to Anomalies

The lesson for today's projects is that, in a critical situation, any anomalies spotted are acted on, with a smooth escalation between frontline (lookouts) and tertiary support (officers on the bridge), as in Figure 6.8. This trouble-free escalation needs to be established in the project testing phase and verified through the operability and operational testing. As frontline operations become familiar with the solution and environment, they can then set up procedures that are more effective.

Preparing Operations through Scenarios

The lesson for today's projects is that the operation and technical support staff need time to map critical scenarios for the solution's operability, work out strategies for failure prevention, and determine preset and proven action. These need to be carefully carried out and tested before implementation, and might include considering how automated operators are overridden, otherwise, they could cause more problems in a critical situation. After all, the goal for the project is a successful implementation and prevention of loss of service.

Chapter Wrap-up

Conclusion

At this point, the predicament the officers found themselves in was evidently down to serious deficiencies in the project's earlier phases. For example, time set aside for testing was too short and the officers did not go through any S-turn maneuvers during sea trials, did not simulate handling the ship under rough or dire conditions, and did not simulate an emergency as part of accident prevention.

Many factors contributed to the collision. The ship's officers failed to piece together the ice field's extent and understand the true danger ahead. No attempt was made to slow the ship. There was a serious oversight in how the radio operators reported into the ship's hierarchy, leading to a serious failure in discerning and reacting to critical incoming messages, especially from *Californian*. There was friction between the lookouts and officers over the missing binoculars. The lookouts, without vital equipment, hesitated, which cost vital seconds. Murdoch's maneuver was well executed, but perhaps with longer sea trials and testing, he could have pulled it off better. In hindsight, when problems occurred, the captain and officers should have done more to clarify the scope of anomalies brought to their attention, investigate them more closely, and piece together all the necessary intelligence.

Key Lessons for Today

In today's projects, a project manager preparing for this phase would work closely with the operations team to do the following:

- Ensure the operations team is brought early into the project, and it takes a prominent and lead role in the planning and testing phases. This team is ultimately responsible for upholding the service levels to the business.

- Ensure that both the business groups and the operations team refine the SLAs together.

- Ensure there is a tiered (three or four) support structure for dealing with problem management with preferably a central help desk.

- Ensure operations team has one holistic customer view of the solution.

- Avoid dispersing support staff into technology "silos"; instead, all are available, and they can bring many backgrounds together when solving a problem. Look closely at the overall support, especially, the interface by different support groups.

- Ensure the problem-management processes are based on speed-of-recovery.

- Ensure proactive problem-avoidance is based on an early warning system.

- Ensure information from feedback mechanisms is synthesized and routed timely to decision-makers.

- Ensure an adequate operation (people, processes, tools) is set up around a solution. Otherwise, it will inevitably lead to operational problems that manifest themselves days, weeks, or even months after going live and a potential failure or outage.

Educators

Discussion points:

- Should White Star have reviewed the wireless operation more closely for coverage (were two radio operators enough) based on the traffic volume? Was the process the radio operators followed understood; were the interactions between them and officers clear?

- Did White Star set up operations too late, where officers were introduced at sea trials, and the crew brought in a few days before the maiden voyage?

- How much did overconfidence play into accepting the situation as normal, and not looking more closely at the anomalies?

CHAPTER 9

Collision

In This Time Frame

- April 14, 1912, 11:41 p.m.—*Titanic* strikes ice.
- April 14, 1912, 11:42 p.m.—Two damage assessment parties are sent out.
- April 14, 1912, 11:50 p.m.—The first damage assessment party returns.

Overview

This chapter looks at the collision in some detail—how passengers, officers, and crew described it. It also looks at the immediate impact, the disposition of damage assessment parties, and the subsequent sequence of events that occurred on the bridge. After the collision, the ship appeared to be in remarkably good condition. No one had been injured, and from the bridge, the ship's integrity seemed to be sound. However, a completely manageable situation was turned into a catastrophe, as business pressures overrode standard procedures.

The Collision

Titanic's officers tried desperately to avoid a collision; however, the S-turn, (Figure 8.4) a good decision, failed to decelerate the ship enough, so it struck ice.

Description of the Collision

At 11:41 p.m., *Titanic* halted, later described by hundreds of passengers as a quiver, rumble, or grinding noise, as if the ship were rolling over a thousand marbles. The consistent testimonies of the collision described it as almost innocuous. Here are a few:

> *"My mother felt a little bump... it wasn't enough to awaken anyone."*
>
> —Eva Hart (born 1905, recent Interview)

> *"I heard this thump, then I could feel the boat quiver and could feel a sort of rumbling."*
>
> —Joseph Scarott, Seaman

> *"... It was like a heavy vibration. It was not a violent shock."*
>
> —Walter Brice, Able-Bodied Seaman

> *"...I felt as though a heavy wave had struck our ship. She quivered under it somewhat."*
>
> —Major Arthur Peuchen, First-Class Passenger

> *"I was dreaming, and I woke up when I heard a slight crash. I paid no attention to it until the engines stopped."*
>
> —C. E. Henry Stengel, First-Class Passenger

> *"We were thrown from the bench on which we were sitting. The shock was accompanied by a grinding noise...."*
>
> —Edward Dorking, Third-Class Passenger

In short, most survivors recalled the collision as a slight tremble of the ship, in which the boat quivered for not more than ten seconds.

Collision Leaves Titanic Grounded

Through the collision, there was no "crash stop" (the propulsion machinery is set to full-astern), fatalities, or even minor injuries. There was no violent jolt sideways or repeated strikes along the ship's length. This is common with a sideswipe against an ice spur when a ship turns very hard away from it. This would have created chaos in a ship sailing at 22.25 knots. Everything in the forward third of the ship would have jumped sideways in a rebound effect, similar to a major earthquake. None of the seven hundred survivors recounted this as a dramatic event, except the men in the stoke-holds (bottom of the forward compartments). The breakfast cutlery that was laid out in the dining salons barely trembled, and drinks remained unspilled in the first class smoking rooms and lounges. All the evidence indicates that the ship came to rest on an underwater ice shelf at the base of the iceberg. Murdoch had prevented a head-on crash that could have demolished the first four compartments and killed or maimed hundreds of passengers.

"...It seemed almost as if she might clear it, but I suppose there was ice under water."

—Lookout Reginald Lee, Crow's Nest

"I was just sitting on the bed, just ready to turn the lights out. It did not seem to me that there was any very great impact at all. It was just as though we went over a thousand marbles. There was nothing terrifying about it..."

—Mrs. J. Stuart White, Passenger, Cabin C-32

"What awakened me was a grinding sound on her bottom. I thought at first she had lost her anchor and chain, and it was running along her bottom."

—Lookout George Symons, Crew's Quarters ("up forward")

"A noise; I thought the ship was coming to anchor."
—3rd Officer Pitman, Officers' Quarters

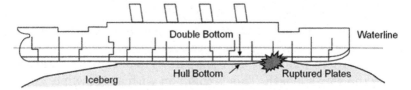

Double Bottom | Waterline
Iceberg | Hull Bottom | Ruptured Plates

Figure 9.1. Titanic ground to a stop on the iceberg shelf, which ruptured through the hull bottom and the double bottom. However, the rate of flooding was contained by the water pumps that could keep pace with a maximum inflow rate of 15 cubic feet per second. Theoretically, this would have kept her afloat for a week.

Preparing for Loss of Service

The lesson for today's projects is that when a solution falters in production, steps are taken to recover it according to a process prepared, planned, and tested in the project itself. The process should be based on a **Mean Time to Recovery (MTTR)** or recovery clock (Figure 9.2), where the operations team's principal objective is to get the solution back in service as quickly as possible to meet Service-Level Agreements (SLAs). The solution is then patched up in the background and a temporary or permanent fix applied. However, before going back into service, the solution's integrity needs to be first established with a level of testing. Hence, the importance of having the operations team involved in the testing phase, so the problem does not recur. With an eye on the clock, the operations team steps their way through the process and the four "problem" quadrants of detection, determination, resolution, and recovery. When the recovery clock starts ticking, signifying the loss of service starting point, metrics should be captured as User Outage Minutes (UOMs) because that measures how many users experience service loss and for how long.

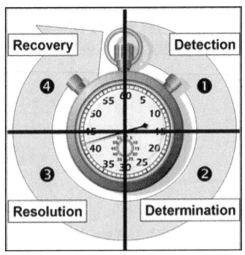

Figure 9.2. The four problem-recovery quadrants of detection, determination, resolution, and recovery. The model/should be verified in the project testing phase.

If we analyze the *Titanic* story through this problem-recovery, starting with the **problem detection quadrant,** the *Titanic* lookouts gave 37 seconds of warning. Unlike today, this is not typical with a solution, which is likely to put out errors and warnings well before any significant failure occurs, providing operators, including automated systems, time to prevent the problem from occurring.

Gathering Preliminary Information on the Damage

Grounding along an underwater ice shelf greatly damaged the outer shell and the internal frames. The impact would likely have crushed the double bottom, although inconsequential, as it opened the ballast tanks to the sea beneath the watertight tank top deck. Within seconds of the collision, some flooding had occurred in the coal bunkers and Boiler Room 5. One fireman later testified seeing a gaping hole two feet into the coal bunker's floor. Suction lines were set up right away, and the water pumps coped with the rate of flooding, keeping the ship afloat. Meanwhile, on the bridge, *Titanic's* captain, director, and officers gathered (Figure 9.3) to determine action.

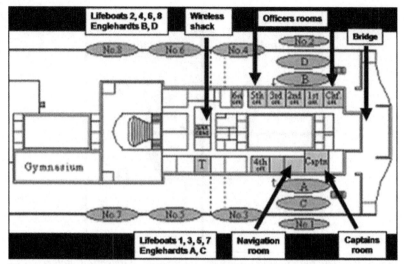

Figure 9.3. In this layout of Titanic's top deck and bridge, note the ease of access to the bridge or control center from the captain's and officers' quarters, allowing quick decision making.

At 11:42 p.m., as part of standard naval procedures (problem determination quadrant), two damage assessment parties were sent into the ship's bowels to determine the extent of the damage, front and mid-ship.

At 11:50 p.m., the first damage assessment party returned, and Fourth Officer Boxhall delivered a positive report to Captain Smith that they had found no major damage or flooding, and everything appeared in good condition.

Deciding Action

Bruce Ismay and Captain Smith conferred to determine action, and they were boosted by the preliminary report that perhaps the situation was not as bad as first feared. Unfortunately, in Bruce Ismay's mind, the assessment (**problem detection and determination quadrants**) was now complete. A more thorough problem determination was not needed. Bruce Ismay now looked for a resolution (**problem resolution quadrant**).

Understanding the Root Cause

The **problem determination quadrant** is completed when the problem's cause assumptions are tested and proved correct. Today, not working out the precise sequence of events before a solution failure could lead to a misdiagnosis where a wrong fix is applied, the resolution fails, and the problem recurs. Therefore, it is important to be sure of the evidence at hand and to ask the following:

- Was there an awareness that the solution was going to fail? If so, were any (automated) preventive actions tried?

- Did the solution alert human or automated operators?

- Were any feedback mechanisms faulty and providing unreliable data?

- Is the diagnosis of the problem correct?

Alternative Resolutions for Bruce Ismay

So, what action was available to them? Resolution with a distress call was a concern for Ismay, as it would compromise White Star's position by shattering the hype around the Olympic-class ships. It would destroy the brilliant marketing that had lured the world's wealthy elite onto the safest liner ever built.

A better resolution would be to get the ship back to Halifax, away from New York and the center of the world's press. Bruce Ismay could then better contain the news story and marginalize it as a minor incident. Passengers would be disembarked onto trains, and the ship would be patched up and sailed back to Belfast for repairs. Bruce Ismay could boldly claim that *Titanic*, a lifeboat in itself with all the latest in emerging technologies, could save herself from a potential disaster and further push the safety claims of White Star lines. This would be a major publicity coup, on a much greater scale than *Arizona* that collided with an iceberg and then saw a sharp rise in ticket sales because she was perceived a far safer ship than wooden ships of the day. From a business perspective, this alternative was very attractive for Bruce Ismay. From an operational perspective, this had many risks. For example, what would be the outcome of moving the ship off the ice? Could this cause further damage?

Recovering Loss of Service

The lesson for today's projects is that when a solution falters the **problem determination quadrant** assesses the impact on users. Problem determination has to be consistent with all the available evidence, and all the anomalies are investigated. This includes the reinvestigation of feedback mechanisms and logs important to determine whether the problem has been building up and its cause.

Turning a Critical into a Catastrophic Situation

Titanic's situation was critical, but not catastrophic. Bruce Ismay was hellbent on saving face, and his anxiety over White Star's reputation created an atmosphere where mistakes were easily made. *Titanic* appeared to be completely stable, sitting snugly on the underwater ice shelf. Perhaps, with due care, they could dislodge the ship with minimum damage. Bruce Ismay rushed into making a decision. The second assessment party, with the architect and carpenter, had not even returned with an assessment that would be better qualified. At 11:50 p.m., 10 minutes after the collision, Bruce Ismay pushed to restart the ship and limp *Titanic* off the ice shelf with the purpose of sailing to Halifax. Captain Smith made the fateful decision to sail forward and telegraphed the engine room "dead slow ahead" to recover the situation. Engineers later testified the ship moved forward at 3 knots with a grinding noise

.

Chapter Wrap-up

Conclusion

The innocuous collision created little collateral damage and failed to raise any serious concerns, although damage assessment parties were sent out. The first damage assessment party returned and just reinforced this message. Whether Captain Smith was part of the decision to restart *Titanic* was irrelevant at this point, as Bruce Ismay controlled the situation driving forward his agenda. With Bruce Ismay in control, he could influence the actions, and as a result, not all courses were adequately explored as part of the problem resolution.

Key Lessons for Today

In today's projects, a project manager preparing for this phase would work closely with the operations team to do the following:

- Adequately plan for a process to deal with problems around a recovery clock to enable the operations team to restore service quickly and maintain service levels.

- Ensure the process carries the checks and balances (through reviews) to minimize the likelihood of mistakes made in a pressure situation.

- Ensure the process outlines roles and responsibilities to ensure the right personnel make the right decisions. More senior people in the hierarchy not responsible for the area cannot overstep it.

- Predefine the list of potential scenarios (in the project design phase as outputs of static testing) to help where the time to respond is very short.

- Consider alternative actions available, with the risks associated with each action. Only then, should the last quadrant of recovery begin, and the operations team restores the service, according to SLAs.

- Assess the overall risks before applying a resolution or fix to production.

- Ensure the process can handle executive intervention and ensure it stands up to careful examination not to deteriorate the situation further at all. Importantly, it needs to be challenged without repercussions if it does not make sense. The operational team needs to be protected by governance from stakeholder (or sponsor) meddling.

- Verify in the testing phase the operational procedures for the solution and the way they are performed, through an assessment of operational readiness.

Educators

Discussion points:

- Discuss the likely scenario on the bridge minutes after the collision and review the actions available to Bruce Ismay and Captain Smith.

- What sort of governance could be in place to shelter the operations team from stakeholder (or sponsor) meddling and protect the operation's integrity?

- Under what sort of pressure was Captain Smith? If the situation was stable, could he not have waited and delayed his decision.

- Could Captain Smith's pending retirement account for his passive approach?

CHAPTER 10

Restarting Engines and Sailing Again

In This Time Frame

○ April 14, 1912, 11:50 p.m.—Engines restarted and *Titanic* taken off ice.

○ April 15, 1912, 12:40 a.m.—*Titanic* is sinking (Captain Smith issues orders to uncover lifeboats).

Overview

This chapter looks at the sequence of events after the collision that led to restarting and moving the ship off the ice. Only after the second damage assessment party returned, with the ship's architect (Andrews) and carpenter, was there full realization of the consequences of Bruce Ismay's actions.

Everything Back to Normal, or Was It?

At 11:50 p.m., Captain Smith proceeded to the wireless room to inform the White Star Line in Boston of the situation. Captain Smith was still optimistic; after all, there was great confidence in the ship's design with 73 watertight compartments. Captain Smith sent a wireless message outlining that

Titanic had struck ice but with little damage. Everyone was safe aboard, and as a precaution, the ship was proceeding to Halifax. The message would give White Star time to organize trains and carriages to transport the passengers to New York. Wireless messages were not encrypted, and the world media intercepted this one. This was the reason early reports of the collision that appeared in the European press were overwhelmingly optimistic.

Passengers, unaware of any dangers, later testified their initial relief that the ship was restarting the journey, with little concern about the collision, the potential damage, and any consequences.

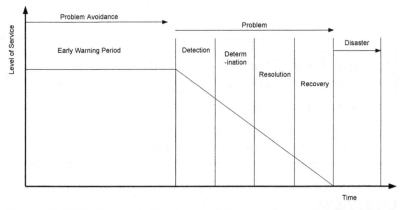

Figure 10.1. The first period in the graph is an early warning period. This is extremely important for problem avoidance as it signals an impending problem is about to happen. The project team in the project testing phase should verify that operations use a similar type of model.

In today's projects, recovery procedures need to be controlled by the operations team who are responsible for the services they provide. Any communications or announcements related to a service outage need to be made with their blessing and participation, and full support given to the decisions made. This is critical when communicating externally with service recipients, as inaccurate information can quickly erode confidence in the service provider.

Second Damage Assessment Party Returns

At midnight, the second damage assessment party returned with the architect Thomas Andrews and the carpenter John Hutchinson. They had a more accurate assessment of the situation and better data. The first damage assessment party had not descended enough decks to see the damage's full extent. Andrews' team had, and he knew from the ship's design, that if the mail room was lost to flooding, then the ship was doomed.

The lesson from this for projects today is that to pinpoint faults, the operations (support) team needs detailed knowledge of the solution and the way it is integrated with the environment and the ability to decompose and analyze it. Documentation must be created at each phase of the project and then transferred as knowledge to the support staff for use in the operation.

Flooding Becomes Catastrophic

At 12:10 a.m., 20 minutes after restarting the ship, Boiler Room 6 had started to flood. Apparently, the initial determination was grossly inaccurate, and Bruce Ismay's decision had turned a bad situation into a catastrophic one. The mail room was soon lost to flooding. Captain Smith conferred with Andrews and the officers, determined that the ship—sailing now at 8 knots—should gradually stop.

In a recovery situation where a solution falters, it is important to keep assessing and reassessing the environmental data (evidence) and monitoring the environment for any changes. The first fix applied is usually temporary to get service back up as quickly as possible. It might take hours or even days to get a permanent fix in place. This then needs to go through rigorous planning and testing before implementing the changes into production using procedures developed during the project. Hence, a robust change management process and a test/staging environment are required.

The Architect Predicts a Two-Hour Window

The forward motion had taken its toll, and the ship had taken on more water. Parts of the ship initially unaffected had started to spring leaks, and the increase in flooding became catastrophic. Andrews correctly predicted to Captain Smith that the ship had about two hours before foundering. This was a death sentence, and Captain Smith finally recognized the situation as hopeless—not recoverable as it had been right after the collision. Bruce Ismay had forced a situation where the ship went beyond the possibility of recovery.

The lesson from this for projects today is that recovery procedures can usually allow several recovery attempts within a limited period. The priority should be getting the service back up, and often a temporary fix is applied while a permanent fix is created. In such a situation, it is essential that the solution be closely monitored to determine whether the fix works and holds up.

Damage Report

The second damage assessment party reported major flooding in five compartments and recognized that *Titanic* was not designed for this. The grinding along the bottom had very much ruptured the outer skin and damaged the double hull. The different rates of flooding in the six primary compartments indicated the top hull (or tank top) was damaged. It was beyond the designer's expectations that something in nature could inflict so much damage.

The lesson for today's projects is it is important that a project team plan for such an eventuality, where the fix does not resolve the problem, and the situation goes beyond recovery. The service is unavailable to end-users and customers and is not readily recoverable anymore. For this situation, disaster recovery procedures need to be set up, prepared, planned, and tested by the project team and institutionalized with the operations or technical support staff.

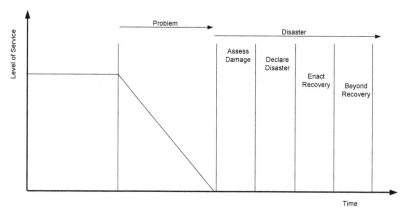

Figure 10.2. The model shows the deterioration of a situation that takes it from a problem to a disaster. The operations team should verify the model in the project testing phase.

Lost Situation

Thomas Andrews was one of the first on board to realize the situation had gone beyond normal problem recovery and had become a disaster. He accurately determined that the problem could not be fixed. Too many compartments were ruptured, and they were rapidly flooding beyond the capacity of all the pumps. The bulkhead walls, separating the compartments, had not been carried up to watertight horizontal traverses. Therefore, as the ship's nose went down, water spilled from one compartment to another rather like an ice cube tray filling with water (Figure 10.3). The ballroom acted as a massive channel for distributing water horizontally across the ship.

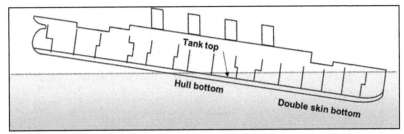

Figure 10.3. All the ship's safety functions were rendered ineffective, except the lifeboats, as water spilled over the bulkheads from compartment to compartment.

Uncover the Lifeboats

Only Captain Smith and a few officers knew the damage's extent, and they were now resigned to the ship sinking. No "abandon ship" command or formal declaration of a disaster was given. About 65 minutes after the collision, Captain Smith gave orders to the officers to uncover the lifeboats and get the passengers and crew ready on deck. No formalized disaster recovery plan was in place on board *Titanic*.

Poor Communication

Titanic's captain understood the situation's seriousness quickly after the collision, but did not tell either the ranks of crew or the passengers on board. This increased the confusion, particularly with the crew. For example, the engine room sent some engineers to the boat deck, but the bridge sent them back down to the engine room. There are several possible explanations for the poor communication aboard *Titanic*:

- The ship had very limited communication capabilities, with no public-address system. Important information was told to passengers by word of mouth, the crew knocking on each cabin door, and making announcements in the common rooms. This would have taken hours, considering there were hundreds of cabins.

- The crew didn't have accurate information on the situation, so varying degrees of information were passed to passengers. The experienced Captain Smith believed in the ship's safety systems and might have found the architect's verdict very hard to accept because everything seemed so normal in the first hour. Captain Smith acted as if the situation were "business as usual."

- Captain Smith realized that the carrying capacity of the lifeboats was inadequate, with only enough room for about half the estimated 2,223 people on board. There were no mass communications to keep things calm and to allow the lifeboats to be filled orderly when the timing was right. The ship's hierarchical structure and segregation of classes meant that first-class passengers had the best access to the boats.

- Captain Smith feared widespread panic. He and the other officers knew what had happened aboard the French liner *La Bourgogne* that sank 14 years earlier. With room in the lifeboats for only half the people on board, widespread panic could have broken out. Captain Smith knew he could save the maximum number of lives by loading only those fortunate enough to reach the boats. Therefore, he might have avoided informing all the passengers, specifically those in third class.

Chapter Wrap-up

Conclusion

At this point in the story, we see how the compromises to the nonfunctional requirements during the construction phase of the project had a major consequence in the disaster. Although, in hindsight, this was a risk White Star was willing to accept.

Today, many projects compromise the post-implementation period by not adequately preparing an operations team for worst-case scenarios and then aligning these to recovery procedures. Today, recovery procedures are not enough and require a well-thought-out communication plan. The operations team needs to be not just aware, but also trained in these.

Many critical situations are compromised by not following an established recovery procedure for a solution. Institutionalized recovery procedures help reduce the risk of something going wrong by minimizing disparate decision making, as was carried out by Bruce Ismay, and prevent a critical situation from becoming catastrophic. In such situations, it is critical to use the operations team's (support staff) detailed knowledge of the solution and its behaviors, as this reduces the risk of a faulty recovery.

Key Lessons for Today

Today, a project manager would work typically closely with the operations team to do the following:

- Review current disaster recovery plans and determine whether adequate to recover the solution.

- Review the recovery procedures and their sign-off on the implementation of a fix, and determine how high up the organizational chain the sign-off is given.

- Every disaster recovery plan needs to be accompanied with a well-thought-out communication plan that clearly communicates with different audiences. The communication plan is probably as important as a disaster recovery plan for several reasons:

- ○ Communicating internally with employees can greatly help control the impact of a disaster. The speed of communication is also essential. For example, get information to customer-facing employees first, so they can inform customers.

- ○ Communicating externally with customers is essential, and the plan needs to cater to customer segments using different channels, depending on the scope of the problem or disaster. A customer-retention strategy might need to be offered.

- ○ Communicating with the public media or press might be necessary depending on the severity of the loss of service. This requires the identification of key messages, how these are communicated, and through what channels. Many companies have been caught off guard when roving reporters trap unaware employees with questions.

Educators

Discussion points:

- • Should executives be involved in the sign-off on the recovery of a solution for which they are responsible? Could there be a conflict of interest?

- • Was Captain Smith's reluctance to declare a disaster situation with an "abandon ship" order justified?

CHAPTER 11

The Disaster Unfolds

In This Time Frame

- ○ April 15, 1912, 12:24 a.m.—Lifeboats readied for lowering.

- ○ April 15, 1912, 12:45 a.m.—First lifeboat launched with only 27 people (65 seats).

- ○ April 15, 1912, 12:45 a.m.—Disaster signs are now obvious to all on board.

- ○ April 15, 1912, 2:30am—*Titanic* finally sinks.

Overview

This chapter looks at how the disaster unfolded. Although risks had been identified and envisioned, and rescue and recovery scenarios had been played out earlier in the project, the recovery was so poorly enacted that six of the first eight lifeboats left half-empty.

What Is Disaster Recovery?

Today, disaster recovery is the concept of switching the operation to an alternative. It takes many shapes and forms, from the relatively simple recovery of data and files in a period measured in days to the complex recovery of an enterprise business solution in a period measured in minutes or hours. The latter is also called business continuity. A disaster can take three forms, namely: (1) total (absolute and immediate), (2) rapid and imminent, (3) slow and innocuous. When a disaster is recognized, contingency plans should be invoked and a disaster declared.

Envisioned Rescue and Recovery Scenarios

On board *Titanic*, for most passengers, the disaster unfolded slowly and almost innocuously, visible only in the lower decks. Although a full recovery was not feasible anymore, the opportunity was lost with Bruce Ismay's decision to come off the ice. Captain Smith and officers enacted a partial recovery. The best they could do was to bring some order to prevent widespread panic and chaos once the disaster signs became obvious.

The envisioned scenario for disaster recovery, in the design phase of the project, was to transfer passengers through lifeboats to another ship and then deliver them to port. The lifeboats would ferry passengers back and forth to the rescue ship therefore requiring a much smaller total lifeboat capacity than a lifeboat place for everyone would have given. This scenario was based on the perception that *Titanic* was unlikely to sink, but would float incapacitated waiting for help. It also assumed that the North Atlantic sea lanes carried much sea traffic, so a ship was always just a few hours away. *Californian* was a few hours away but had pulled up for the night, as it was too dangerous to sail through the ice-strewn waters.

Defining Types of Failures

Today, in defining a (disaster) recovery plan for a solution, three types of failures could happen:

- Design errors that lead to design failures
- Physical faults or failures in the technology

- Operations errors caused by staff due to accidents, inexperience, lack of due diligence or training, not following procedures, or even malice.

External factors such as natural disasters and terrorist activities can be equally devastating. These typically are taken care of by the overall disaster recovery plan for the whole operation.

Risk Identification

For the past 400 years, most risks related to crossing the Atlantic had been observed, identified, and documented. White Star was very familiar with these risks, which included year-round natural conditions, such as changing ocean currents, and weather patterns, such as storms and hurricanes; natural hazards such as fog banks, ice fields, and iceberg areas, and dangerous shorelines, rocky outcrops, and similar features. Through the project a belief had evolved that all these risks, anything nature could hand out, could be handled by the ship's design.

Defining a Disaster Recovery Plan

Today, in defining a disaster recovery plan, the scale of disaster is also important to consider. For example, if a relatively minor storm, fire, or flood knocks out an operation, customers will expect some service contingency plan to kick in quickly. Today, contingency is needed for all these, even catastrophes. The associated costs of disaster recovery vary, based on the window of recovery (time), the elements of the disaster, and the degree of recovery required. As part of the plan, these costs, especially for the solution, need to be determined in the early phases of the project in the business case.

White Star's Recovery Plan

For White Star, under maritime convention, a simple recovery plan was defined for all the above situations. Such a plan would have brought everyone on board to the boat deck, loaded them into the lifeboats, lowered the lifeboats safely, and put them adrift with the experienced crews to handle them. The lifeboats would ferry passengers back and forth to a rescue ship, which would then deliver them to port. The lifeboat drill in Queenstown should have tested how the lifeboat crews not just lowered the lifeboats, but handled them in the water.

How Solutions Fail

Today, many serious problems with a newly implemented solution can start so innocuously that, in the first hour, operations might not even know the problems or their implications. For example, a less critical part of the solution might be "down," so it goes unnoticed. However, because of dependencies between parts, a "domino effect" can quickly affect other parts of the solution, which can lead to a catastrophic failure very quickly. Hence, in the project design phase, it is important to establish the integration requirements between the proposed solution and the existing services (environment), with any dependencies (see nonfunctional requirements in chapter 3). This all then needs to be scrutinized in the testing phase, through various scenarios.

Delay in Launching the Lifeboats

On board *Titanic*, there was a major delay in getting the lifeboats down, indicating a hesitation to launch the boats until as late as possible. The officers likely reacted slowly for several reasons: the ship was believed to be unsinkable; the gravity of the situation was not apparent; and everything seemed normal at the time. In addition, only 83 of the crew of 900 were mariners and knew the somewhat complex drill of lowering a 30-foot (65-person) lifeboat 60 feet to the water. There were 16 of these lifeboats, plus four smaller collapsible (45-person) lifeboats, or "Englehardts."

Skepticism Reigns

The crew carried out orders to launch the lifeboats, but for a long time, they were skeptical that anything serious would happen to *Titanic*. Most passengers were unaware of the disaster, and the lifeboats were filled on a first-come, first-served basis from the top decks, mainly first and second-class passengers. Although each lifeboat had a capacity of 65 people, only 27 very reluctant people were lowered in the first at 12:45, about 65 minutes after the collision.

Business Continuity

Today's organizations cannot afford a project failure on the scale of *Titanic*, hence, the importance of developing plans in the project design phase that manage the risk and reduce the risk of failure. Sometimes, these critical plans look post-disaster at how an organization continues to function. Business continuity is an advanced approach to disaster recovery that goes beyond just the technology to recover the whole business operation. This requires replicating the business operation, including the staff, all the processes and procedures, hard copy records, and documentation.

Launching of Lifeboats Begins

From the first lifeboat's launch, *Titanic's* recovery window was just more than one hour, during which all lifeboats had to be (Figure 11.1) filled and lowered, a stretch for the small crew of 83 mariners. However, the disaster was not imagined by the architects who believed the Olympic-class ships would continue to float.

Figure 11.1. All the lifeboats were on the boat deck (top deck) with a very long and dangerous drop to the waterline (60 feet).[1]

Estimating Potential Revenue Losses in a Disaster

Today, the disaster recovery time or recovery window determines the disaster costs, as **revenue loss is usually directly proportional to the recovery time**. It is important to assess the downtime cost of the solution and determine whether the business could slip into a serious financial situation, or even bankruptcy. The recovery window is the major variable in disaster recovery planning, and it should be based on the greatest window the business can possibly afford.

Role of the Bridge

On board *Titanic*, the bridge was the command post. As a result, all major decisions and actions continued to be directed from this location, for example, directing appeals for help through wireless distress calls, slowing the flooding by shutting watertight doors, and preparing the rockets for firing. At this point, the first of eight distress rockets was fired.

Directing the Recovery Operation

Today, a business continuity plan outlines a disaster recovery method that starts with setting up a command post to direct the recovery operation. The plan also includes the procedures for restoring the business services and functions, difficult because of the nuances and anomalies of each organization. The plan defines responsibilities and guides those executing it.

Lifeboats Leave Half-Empty

As the first few lifeboats were lowered, the extent of the damage was still not very visible. The sea was calm, and a slight list of five degrees was usual during the crossing. On board, the perception was that the ship could withstand collisions with a dozen icebergs and stay afloat. Passengers had no preparation for this disaster, as no lifeboat drills had been practiced during the voyage. Many passengers arrived on deck but wandered back to their warm cabins. It took much persuasion to get the women passengers into the cold lifeboats, leaving their husbands and the ship's warmth behind. The next three lifeboats left with only 28, 41, and 29 passengers, respectively.

Preparing the Business People

Today, in the project design phase, it is important to pay attention to the business procedures around a solution and work with the individual business units and groups to develop contingency plans for these. This dramatically increases ownership and the ability to respond in critical situations.

Poorly Enacted Recovery

After 1:00 a.m., the warning signs became obvious. Third-class passengers and stokers who had managed to get from the decks below had seen the seawater rising up the floors firsthand. They were ready to fill the lifeboats. Panic did not break out on *Titanic*, however; passengers for the most part filed orderly into the lifeboats.

Boat #	Time of Launch	Total People
6	12:55	28
8	1:10	29
10	1:20	55
12	1:25	42
14	1:30	63
16	1:35	56
2	1:45	26*
4	1:55	40
D	2:05	44
B	Floated off	

Boat #	Time of Launch	Total People
7	12:45	27
5	12:55	41
3	1:00	50
1	1:00	12
9	1:20	56
13	1:30	70
15	1:35	64
C	1:40	71
11	1:45	70
A	Floated off	

Figure 11.2. The sequence of lifeboats launched on port and starboard is very telling. Six of the first eight lifeboats left half-empty. The last two Englehardts were floated off upside down.

Developing and Testing a Business Continuity Plan

Today, a developed business continuity plan not only lists all the procedures in a "run book" but also the daily activities, tasks, and the required information flow. It is important to perform a dry run and test this plan thoroughly in the project testing phase just to see how people react to it and carry it out.

Problems in Lowering the Lifeboats

Lowering a lifeboat under normal conditions was tricky because of the 60-foot drop. If the lifeboats were overloaded, they could buckle in the middle under the weight. By 1:30 a.m., as the ship listed to port, the lifeboats on port swung away from the ship, and getting into these became very hazardous (Figure 11.3). On starboard, the lifeboats swung into the ship and bounced along the ship's side dangerously. There was still hesitation to get in, and at 2:00 a.m., 140 minutes after the collision, the 10th lifeboat was launched—the only full lifeboat.

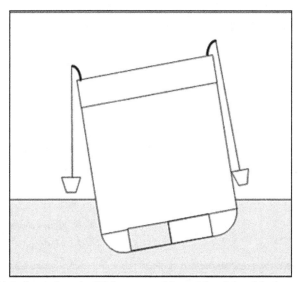

Figure 11.3. Lowering the lifeboats on either side of the ship became more challenging as the ship started to list to port, with the uneven flooding of the cells in the double hull.

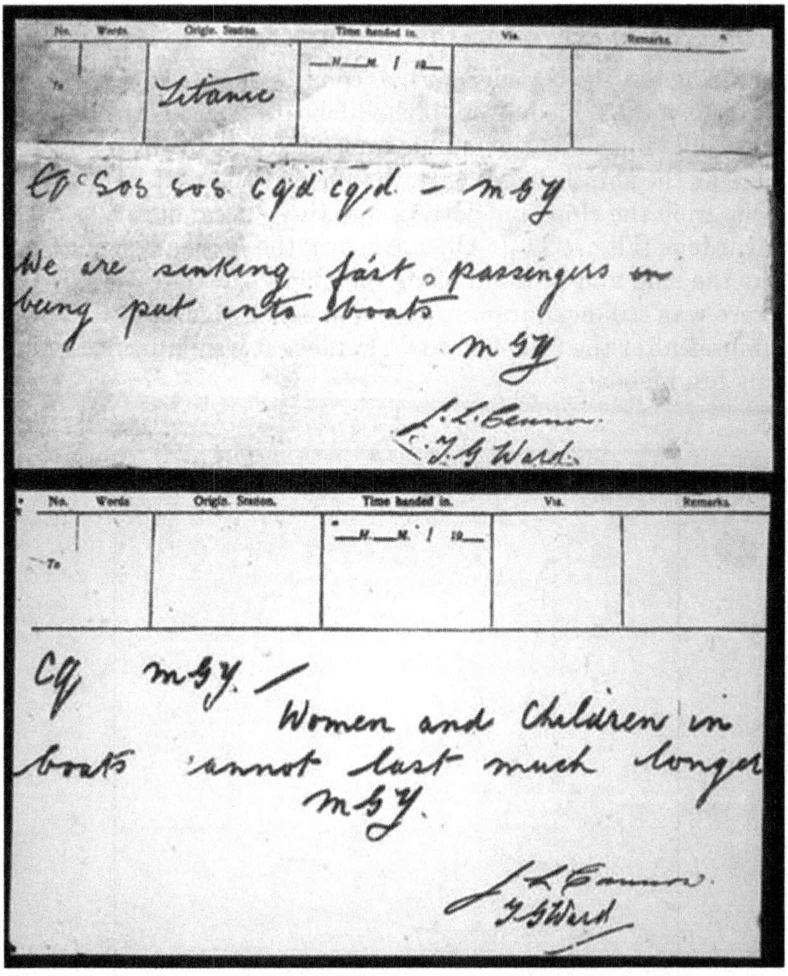

Figure 11.4. These two telegrams highlight the deteriorating situation on board.2

Chapter Wrap-up

Conclusion

Initially, Captain Smith's hesitation to take any action led to a major delay in responding to the disaster, which further propagated problems such as the launch of the lifeboats. The opportunity for a full recovery had gone with Bruce Ismay's decision to restart. The delay led to six of the first eight lifeboats leaving half-empty.

Today, many projects ignore disaster recovery as something beyond their scope and as something covered by a yearly planning process by the IT department. Yet, the project team determines the business justification and the design around the solution, and so, they can best develop in-depth understanding of the kind of recovery required for it. The project is in the best position to identify the business procedures required for the plan. Serious thought needs to be given to the consequences of a disaster affecting the solution, which needs to be done early in the project so adjustments to the overall disaster recovery plan can be made. Recovery should not be an afterthought until the project completes. A disaster recovery plan is worthless unless it has been exercised to check it. Many plans look good on paper but fail in practice due to omissions or false assumptions. At a minimum a desk top exercise should be conducted.

Key Lessons for Today

Today, a project manager would typically:

- Identify and envision potential recovery scenarios for the solution, and the associated risks with these, in the project design phase (through the shipbuilder's model).

- Identify the procedures that should be followed for restoring the solution.

- Ensure that the disaster recovery and business continuity plans are carefully reviewed and how the solution fits into these.

Educators

Discussion points:

- Can the cost of a business continuity plan be justified?

- Should senior officers have challenged Captain Smith over his inability to act, intervened much earlier, and initiated the launch of the lifeboats? Or was the organizational hierarchical structure too inhibiting? Note it was a paramilitary structure.

- What could Captain Smith have done to increase the number of lives saved?

CHAPTER 12

Titanic's Final Minutes

In This Time Frame

○ April 15, 1912, 2:20 a.m.—Last of the lifeboats is floated off.

Overview

This chapter looks at the last few minutes before *Titanic* foundered. Time overcame the recovery to the point that the last two Englehardts were floated off upside down. The officers had to control an explosive situation where some scuffles broke out. As the ship finally sank, lifeboats rowed away for fear of being swamped by survivors. The chapter also speculates what could have happened if a modern disaster recovery plan had been planned for as part of **Risk Management**.

Running Out of Time

Many crew (325) and officers (50) bravely remained at their posts, working in holds, boiler rooms, and throughout the ship, ensuring that electricity was available for as long as possible. Consequently, they did not survive. The launch of 16 lifeboats took more than 90 minutes partly because the order was given so late, and the crew was not very familiar with the drill. There was not enough time to properly launch the last two Englehardts (collapsible lifeboats), which were floated off upside down.

An Explosive Situation

Even if more lifeboats had been available, it is unlikely they would have been launched in the given time. There were only 1,178 lifeboat places and 2,200 people on board. The last 30 minutes were frenetic, as most people realized the full extent of the disaster. The officers manning the launch of the lifeboats were under much pressure, as they controlled an explosive situation. Some scuffles broke out around the last lifeboat, and gunshots were heard. In contrast, there were also many heroic acts, as some wives refused to part from their husbands and get into the lifeboats.

Today, a disaster of this proportion can only be mitigated with a business continuity plan and the appropriate recovery facilities. These should be justified early in the project, and their design should become part of the requirements in **Scope Management**.

Media Coverage

The world press already covering *Titanic's* maiden voyage was quickly alerted to the ship's situation, as can be expected considering the publicity and pomp surrounding the event. The initial reports coming back from the ship were unclear and contradictory. In Europe, the press reflected the initial phases of the disaster and published optimistic accounts that *Titanic* had struck ice, but everyone was safe. This is further evidence that this view prevailed on board *Titanic*, which Bruce Ismay took advantage of and tried to limp the ship to Halifax. The first wireless message after the collision was to the White Star Line offices in New York and Boston supporting this conclusion. As the view of the situation changed on board, the North American press reflected the latter phases of the disaster and published more accurate news, as did the later editions in Europe.

```
        COLLISION WITH ICEBERG - Apr 14 - Lat 41° 46', lon 50°
    14', the British steamer TITANIC collided with an iceberg
    seriously damaging her bow; extent not definitely known.

        Apr 14 - The German steamer AMERIKA reported by radio
    telegraph passing two large icebergs in lat 41° 27', lon 50°
    09',--TITANIC (Dr ss).

        Apr 14 - Lat 42° 06', lon 49° 43', encountered extensive
    field ice and saw seven icebergs of considerable size.--PISA
    (Ger ss).


                    J. J. K N A P P

                        Captain, U. S. Navy/
                            Hydrographer.
```

Figure 12.1. Telegram announcing the collision to the world. This news quickly spread to all ships at sea.[1]

Communication Management

Today, organizations facing serious solution failures would typically go to great lengths to hide these kinds of problems from the public media. After all, bad press can destroy reputations and bankrupt a company. In such a situation, an organization needs a prepared communication management plan to handle all communications between itself and the outside world, customers and media alike. This plan should be created or reviewed (if it exists) during the project as part of the communications strategy. A plan in these situations helps the organization carefully follow a process to respond consistently and address what is important composedly. For example, the public relations department would focus on the messages to the media, while the chief executives would call key customers.

Titanic Finally Disappears

The air temperature was freezing. The sea temperature was below the freezing point, and survival time in these temperatures was 20 to 30 minutes. Everything that might float was thrown overboard, in the hopes of use as floating rafts. On April 15, 1912 at 2:20 a.m., the world's largest ship went down to the bottom of the Atlantic.

Figure 12.2. The view from the lifeboats was chaotic as Titanic finally disappeared beneath the waters. The lifeboats edged away from the center in fear they would be swarmed.[2]

Search for Survivors

Initially, the lifeboats had rowed away from the sinking ship for fear of being swamped by survivors in the water. Later, a few lifeboats returned to the wreck site before *Carpathia* arrived at 3:30 a.m. The lifeboats searched for survivors in the icy water, and amazingly, some were found. *Carpathia* had raced at greatest risk through the ice-strewn waters to get there just more than an hour after the sinking (Figure 12.3). This indicated how much traffic was in these well-traveled sea lanes of the North Atlantic. It gives some credence to the thinking around the project design decisions with the lifeboats, perceived as ferrying vessels rather than life-sustaining vessels that could survive open water for several days.

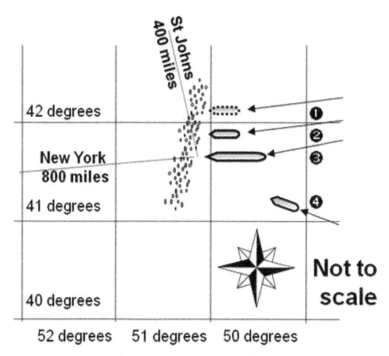

Figure 12.3. Titanic's disaster and rescue scene outlines the position of the disaster and the overall directions from which ships came to the rescue: (1) Californian's position as given at the testimony; (2) Californian's likely position; (3) Titanic's foundering position; and (4) Carpathia's first position, 58 miles away.

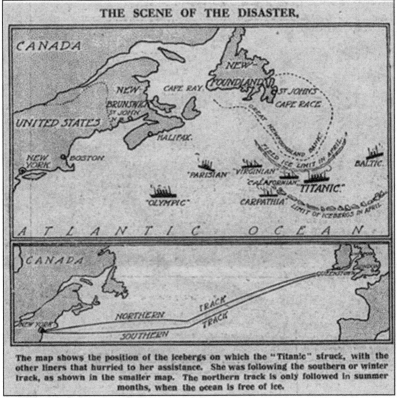

THE SCENE OF THE DISASTER.

The map shows the position of the icebergs on which the "Titanic" struck, with the other liners that hurried to her assistance. She was following the southern or winter track, as shown in the smaller map. The northern track is only followed in summer months, when the ocean is free of ice.

Figure 12.4. The top map outlines other ships near Titanic's disaster site. The bottom map shows the northern and southern tracks for summer and winter months (January to August).[3]

Capturing Lessons for Future Projects

Today, an important element of project management is to incorporate lessons learned into future projects. Let's go back to *Titanic* and speculate what could have happened if a modern disaster recovery plan had been available and had carefully thought out all the possible disaster scenarios. This would have first been identified under **Risk Management**. The shift in thinking, especially for worst-case scenarios, and the overall process undertaken are important in this exercise:

- The first assumption of the White Star planning process would have been the acceptance that the safety systems might fail. This major shift in thinking would build the plan around evacuating passengers and crew off the ship. This would require enough lifeboat seats for all.

- The second assumption would have centered on how quickly the ship could sink and, therefore, the window of evacuation. This would hone the efficiency of the evacuation enacted by the crew within the window. It would require effective lifeboat-launching equipment and a crew trained in its use.

- The third assumption would have considered how quickly a rescue ship could pick up the evacuated passengers and crew from the lifeboats. This would require processes to ensure distress calls were put out quickly. In addition, the lifeboats would have to be seaworthy enough to withstand the harsh conditions of the Atlantic and carry several days of supplies to ensure survival.

The delay in starting the recovery was a considerable factor in the poor way it was enacted and the chaos and danger in the last few minutes before the ship sank.

Figure 12.5. A lifeboat and Englehardt row toward Carpathia. Few full lifeboats were lowered to the sea, adding to the number of people who died during the tragedy.[1]

Figure 12.6. One of the last two surviving Englehardts, lifeboats with collapsible sides, which could accommodate up to 40. Two of these were launched correctly; the other two were floated off upside down, as time ran out.[5]

Figure 12.7. Police stand outside the New York White Star Line offices after the disaster as crowds gather waiting for news.[6]

Chapter Wrap-up

Conclusion

Today, many projects pay little attention to the post-project operation and planning for potential problems or failures. Yet, if a project team considers going through worst-case scenarios early in the project and looking into the future at the dependencies that could grow on the solution, then this would put into context the future value of the solution. It would probably shift the thinking and affect the design process. It would also help with the business justification in the costs of how the solution always needs to be protected.

Key Lessons for Today

Today, a project manager could not do more than already advised in the previous chapter key lessons.

Educators

Discussion points:

- How seriously does an organization treat public media communications during a disaster?

- What is the justification in creating business continuity plans over disaster recovery plans?

- In going through worst-case scenarios for disasters, which of the three assumptions was the most critical?

CHAPTER 13

The (Project) Postmortem

In This Time Frame

- April 15, 3:30 a.m.—*Carpathia* arrives at wreck site.
- April 18, 1912—*Carpathia* arrives in New York.
- April 19, 1912—US inquiry starts.
- April 22, 1912—United Kingdom inquiry announced.
- May 2, 1912—United Kingdom inquiry hearings opened in the Wreck Commissioner's Court in Westminster.

Overview

This chapter looks at the two inquiries (postmortems today) in some detail.

Carpathia Arrives

On April 15 at 3:30 a.m., *Carpathia* arrived at the wreck site and rescued the lifeboats and survivors. *Carpathia* spent four hours at the site searching for survivors.

Figure 13.1. Titanic's survivors are pulled up against Carpathia at dawn on April 15. Note how packed the lifeboat was, likely one of the last ones launched.[1]

Figure 13.2. Titanic's lifeboats were pulled on Carpathia. Captain Rostron ordered a news blackout after the rescue. This was a factor in the US president ordering a full inquiry into the disaster.[2]

Carpathia then sailed carefully through the ice-strewn waters to New York without sending any further news. The world was mystified at this remarkable silence. US President Taft's request by wireless for information was unanswered, so he sent warships to obtain news. The Navy was unsuccessful and believed *Carpathia* was ordered not to answer any queries.

Figure 13.3. New York Times cover story of the Titanic disaster. Newspapers in Europe printed first an optimistic view of the news of the disaster because of the five-hour time difference. The US media printed a more realistic view.[3]

Figure 13.4. The media made much of wireless communications and, in this report, glorified the role of this emerging technology in bringing rescue ships to the disaster site.[1]

Figure 13.5. The London Herald cover story of the Titanic disaster a day after on April 16, 1912, printing a very somber view of the disaster.[5]

Figure 13.6. The Daily Mirror account of the news. The world press jumped on the story, and rumors were slowly replaced with details of the tragedy.[6]

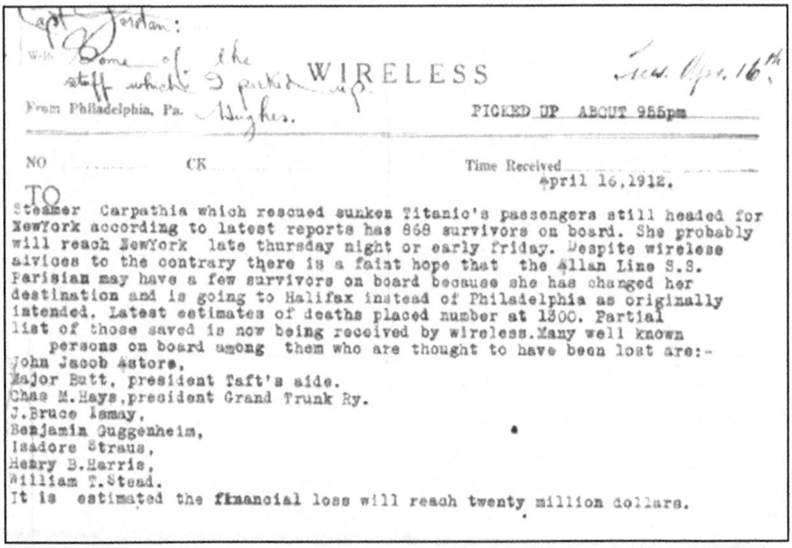

WIRELESS

From Philadelphia, Pa. PICKED UP ABOUT 955pm

NO CK Time Received

April 16,1912.

TO

Steamer Carpathia which rescued sunken Titanic's passengers still headed for NewYork according to latest reports has 868 survivors on board. She probably will reach NewYork late thursday night or early friday. Despite wireless aivices to the contrary there is a faint hope that the Allan Line S.S. Parisian may have a few survivors on board because she has changed her destination and is going to Halifax instead of Philadelphia as originally intended. Latest estimates of deaths placed number at 1800. Partial list of those saved is now being received by wireless.Many well known persons on board among them who are thought to have been lost are:-

John Jacob Astore,
Major Butt, president Taft's aide.
Chas M.Hays,president Grand Trunk Ry.
J.Bruce Ismay,
Benjamin Guggenheim,
Isadore Straus,
Henry B.Harris,
William T.Stead.
It is estimated the fkmancial loss will reach twenty million dollars.

Figure 13.7. The wireless outlined Titanic's survivors were returned to port. Captain Smith had gone down with the ship, although his body was never found. Four officers had survived, and they were critical witnesses in the inquiry.[7] Bruce Ismay, who amazingly had survived the disaster by stepping into the next to last lifeboat as it was lowered, was now the controversial director of White Star. He had hoped for a quick return to England. On board *Carpathia*, Bruce Ismay was given a private cabin where he remained for the voyage to New York. Rumors swirled around the *Carpathia* related to Bruce Ismay. Bruce Ismay was very concerned about the hostile reception he would face in New York. Mr. Franklin received this message from *Carpathia* on April 17:

> "Most desirable *Titanic* crew aboard *Carpathia* should be returned home earliest moment possible. Suggest you hold Cedric, sailing her daylight Friday, unless you see any reason contrary. Propose returning in her myself. Please send outfit of clothes, including shoes, for me to Cedric. Have nothing of my own. Please reply. "
>
> —Bruce Ismay

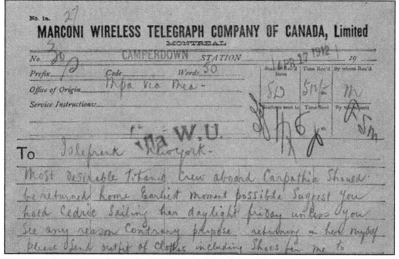

Figure 13.8. The infamous telegram from Bruce Ismay on board
Carpathia to the White Star Line offices:[8]

Figure 13.9. Captain Arthur Rostron ordered that the Carpathia ignore
all media inquiries, he and was criticized by the media for the silence.[9]

Figure 13.10. The world media focus on the disaster was intense. It ran articles on the disaster from when it first happened until the end of the US investigation.[10]

Figure 13.11. Carpathia and the survivors from Titanic. The ship was not very large, and the survivors did not have much space.[11]

Figure 13.12. Crowds in New York City wait anxiously for Carpathia and the survivors from Titanic to arrive.[12]

Figure 13.13. Harold Bride, the surviving wireless operator of Titanic, with his feet bandaged, was carried up the ship's ramp.[13]

On April 18, Thursday, at 8:30 p.m., *Carpathia* arrived in New York where the world's press awaited her—with 1,000 relatives and friends of the passengers.

Today, if you ever face a major operational disaster that affects your customers, you will have to face the scrutiny of the press and your customers' ire. For this possibility, you need a communication plan to handle all communications between your organization and the outside world.

Inquiries Started

Initially, US President Taft intended to do nothing about the disaster. However, the news that the only survivors were those on board the *Carpathia* made the scope of the disaster apparent. Senator Smith asked for passage of his resolution that authorized the Committee on Commerce to investigate the disaster, and the US Senate authorized hearings.

Bruce Ismay Plots His Escape

The Department of the Navy on Thursday 18, 1912, intercepted several significant messages Bruce Ismay sent (see Figure 13.8). These telegrams clearly outlined Bruce Ismay's intent of getting directly back to England, with the remaining officers and crew, without setting foot on US soil.

US Inquiry Is Brought Forward

The Navy contacted Senator Smith, advising him of the intercepted messages, which promoted Senator Smith to set up the hearing to start when *Carpathia* landed, with witnesses being subpoenaed. It started one day after the ship landed on April 19, 1912.

British Inquiry Initiated

The British Government, deeply worried about a US inquiry it could not control, quickly followed suit and ordered its formal inquiry through the Board of Trade. On Monday, April 22, 1912, Sydney Buxton, President of the Board of Trade, requested the Lord Chancellor appoint a Wreck Commissioner to investigate the disaster. The British Wreck Commissioner's Inquiry started on May 2, 1912.[18]

US Outrage over the Disaster

On April 19, 1912, Senator Rayner wrote to the *NY Times*:

> *"Had Titanic been an American ship subject to our criminal procedure, they would be convicted of manslaughter or even murder."*
>
> —Senator Rayner

Sadly enough, Senator Rayner missed that *Titanic* was a US ship. Ironically, it was because White Star was under the US corporation IMM. But because she flew the British Ensign and had British captain, officers, and crew, it was simple to pass *Titanic* off as a British ship.

Figure 13.14. Two views of the rescue ship Carpathia, not a very large ship that rescued more than 700 survivors from the wreck site.[14]

Lives Saved and Lost

The 16 wooden lifeboats (65-person) and the 4 emergency Engelhardt collapsibles (49-person) increased the capacity of the lifeboats—1,186 persons, or 35.5% of total ship's full complement. The total passengers were 2,205, and the lifeboats saved 705 lives, with 1,500 lives Lost.

Figure 13.15. Captain Arthur Rostron and his officers were honored after getting Carpathia back to New York.[15]

Figure 13.16. The media needed a hero, and Captain Arthur Rostron was honored like one. Here, Molly Brown presents him a trophy for his endeavors. Rostron had risked the safety of his ship in his dash to get to Titanic.[16]

The Importance of a Postmortem

Similarly, following a major implementation failure, you need to go through a postmortem of your operation and project. This will focus your organization's learning energies on problem prevention and improve the required service. A postmortem is an investigation of the sequence of events leading up to a failure, or serious problem. It provides insight into the factors contributing to this and helps determine the root causes that otherwise would not surface.

Competition between the Inquiries

Much rivalry and competition existed between the two inquiries as they strove to show which was more thorough. The US inquiry started on Friday, April 19, and the British on May 2. If the inquiry could find gross negligence in how the ship was handled, then White Star faced bankruptcy.

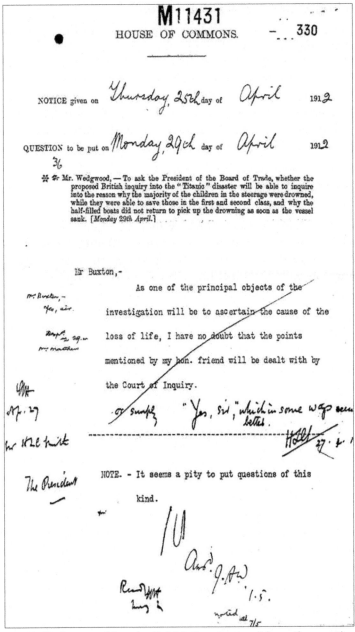

Figure 13.17. *A note asking the British Inquiry to address the question of why most steerage children were lost in the disaster. The public was pushing for answers.*[17]

Postmortem Steps

Typically, there are three steps to take in a postmortem:

- **Step 1: The Discovery**—This examines what went wrong. What should have/have not happened but did/ did not? For this step, you need to collect the evidence (metrics), build the time-line of events, create the problem statement, and determine the factors.

- **Step 2: The Analysis**—This examines why it went wrong. What were the factors? What were the root causes of the factors? For this step, you need to categorize the events and analyze the root causes of the events.

- **Step 3: Corrective Actions**—This examines ways to prevent things from going wrong again. For this step, you need to identify corrective actions to root causes, evaluate your organization's support ability, and implement the changes.

Evidence Collected by Inquiries

As part of the first step, a great detail of evidence was collected from *Titanic's* two investigations. For example, the US inquiry called 82 witnesses within one week of the disaster, including maritime specialists and technical experts. One aim of the US inquiry was to re-create the ship's log through the evidence gathered and to reconstruct the event time-line.

Likewise, in your project, you will need to call witnesses and collect as much evidence as you can through data and metrics. This needs to be done quickly, as it can get lost—or worse, disappear. Pulling this information together is necessary to create the event time-line.

Witnesses Subpoenaed

The US inquiry was worried about losing the opportunity of interviewing witnesses and so had Bruce Ismay and surviving officers subpoenaed. The surviving crew members (200 out of

700) brought to New York were only paid when they were on board a ship and working. Therefore, they were eager to get back to the England and to work.

Figure 13.18. US inquiry chaired by Senator William Alden Smith of Michigan. The inquiry began at the Waldorf-Astoria Hotel in New York the day after the survivors landed.[18]

Required Metrics

Two forms of metrics contribute to an event time-line, namely, hard (or quantitative) and soft (or qualitative) metrics, which can be internal or external to an organization. The operational logs highlight the sequence of events, and they are probably the most important piece of documentation. Typically, these are handwritten, but they can be supplemented with automated tools. Other useful documents include operational run books, end-of-shift reports, standard operating procedures, and procedures related to the project. It is important also to look back to the project and unearth meeting minutes from project core team meetings and executive committee reports (like White Star).

- **Hard metrics** are system-generated and collected through manual or automated processes. Examples include reports, error messages, transactions, and statistics. Uncovering this data requires on-site "archeology." These metrics tend to be explicit, time-stamped, and accurate, but they usually require aggregation. Reports for these metrics include operational logs, service problem reports, service change requests, service-level reports, environmental metric reports, and service outage reports.

- **Soft metrics** include input from the operations team, employees, project core team members, witnesses, partners, customers, and independent bodies. These are solicited through interviews, meetings, and workshops. Soft metrics tend to be subjective, difficult to collect, and prone to error, but they can be very insightful into root causes. They can be collected as witness testimonies, employee feedback, customer feedback, intermediary feedback, marketing feedback, and market perceptions of organization and services. The significance of all these metrics is that they can accurately help piece together the event time-line.

Figure 13.19. Returning crew from Titanic arrive in Southampton.[19]

Bruce Ismay Grilled but Somehow Exonerated

The US Senate inquiry identified that Bruce Ismay was
desperate to get back to the United Kingdom and forced him
to remain in the US, so they could question him over his role
as director and relationship with captain, officer, and crew.
Bruce Ismay was called day 1. But he and the officers had
time to prepare; they were not recalled until day 11, giving
them time to meet and discuss the sequence of events, and
cover-up the truth and what really happened. The stakes
were incredibly high. The testimonies of officers and key crew
members (lookout) were specific to the questions asked; none
offered anything more than was asked. Apparently, they were
well prepared as witnesses or coached into how to avoid saying
anything damaging. On day 11, Bruce Ismay was called to give
testimony.

Figure 13.20. On April 19, 1912, the US inquiry started, and this image clearly carried the date, with, at the center, Bruce Ismay giving evidence on day 1. Inquiries took place on both sides of the Atlantic and uncovered details of the disaster.[20]

Figure 13.21. A second consecutive photograph taken at the US inquiry in the Waldorf Astoria. Bruce Ismay (center) faces the camera.[21]

Contempt of Court

On April 30th, there was talk of civil actions issuing a contempt of court against Bruce Ismay and remaining officers who failed to show up at a hearing.

WASHINGTON, April 30.—Contempt of court proceedings against J. Bruce Ismay. President of the International Mercantile Marine, and Joseph G. Boxhall and Charles Lightoller, surviving officers of the lost Titanic, may be one of the first developments in a civil action begun here to-day to recover for the life of a victim of the disaster.

The three men left the city and went to New York, ignoring summonses issued by the District Supreme Court directing them to appear before a referee and testify in an action brought by Mrs. Louise Robins of New York, the widow of Col. John Jacob Astor's valet, who was lost with the ship. The action is an unusual one in equity to preserve the testimony of the officers for a suit which may be filed later, when they may be beyond the jurisdiction of a United States court.

Ismay, Boxhall, and Lightoller accepted service from a United States Marshal and after a conference with attorneys boarded a train for New York. The hearing, which was awaiting their arrival, was abandoned.

Attorneys for Mrs. Robins say they will ask the court for attachments to arrest the men in New York and return them to Washington to give testimony. Contempt proceedings aginst the attorneys who advised them are threatened.

Subpoenas also have been issued for Frederick Fleet, the Titanic's lookout, and Harold Bride, one of her wireless operators.

Figure 13.22. This article dated April 30, 1912, outlines a contempt of court against Bruce Ismay and remaining officers who failed to show up at a hearing. They ignored summonses issued by the District Supreme Court to testify in a civil action.

Outcome of Both Inquiries

The US inquiry concluded that Captain Smith should have slowed the ship's speed given the icy weather. It called 82 witnesses, including specialists and technical experts. It determined that the ship had reached top speed through the ice field and did not attempt to slow. The US inquiry came close to uncovering the cover-up.

The British inquiry, on the other hand, concluded that maintaining speed in icy weather was common practice.

Both inquiries agreed on who was most at fault—Captain Stanley Lord of the *Californian*. The inquiries stated that if Lord had gone to *Titanic's* assistance when the first rocket was seen, then everyone would have been saved. Both inquiries made recommendations:

- All ships must carry sufficient lifeboats for the number of passengers on board.

- Ship radios should be manned 24 hours a day.

- Regular lifeboat drills should be held.

- Speed should be reduced in ice, fog, or any other areas of possible Danger.

On July 30, 1912, the British inquiry exonerated Bruce Ismay of all blame for the catastrophe. The judgment recommended more watertight compartments in seagoing ships, more efficient drill of the crew, and better lookouts.

Chapter Wrap-up

Conclusion

When projects fail catastrophically, a postmortem is essential to provide a better understanding of the reason for the failure. Every organization needs a postmortem process in place, so little time is wasted in analyzing a significant solution outage or failure.

Titanic's inquiries started within a week of the disaster when the evidence was still fresh, providing us the greatest body of evidence today. The two inquiries failed to uncover the truth, and they were at odds and in competition. The question is was there a cover-up?

Key Lessons for Today

Today, for every project completed, a project manager should do the following:

- Go through a postmortem once the solution has been placed in operation, to capture any lessons learned during the process. This should be wholly removed from whether a disaster or failure happened.

- Set expectations early, at the beginning of the project, about the postmortem to get buy-in, especially from project sponsors. These should be part of any project that meets certain preset criteria (financial, strategic, impact on the business).

Educators

Discussion points:

- Should a postmortem always be completed at the end of a project? What would be an exception to this?

- Discuss the pros and cons of having an external party complete the postmortem.

CHAPTER 14

Was There an Inquiry Conspiracy?

In This Time Frame

o   April 19 to May 25, 1912—US inquiry proceeds.

o   May 2 to July 30, 1912—United Kingdom inquiry proceeds.

Overview

This chapter examines evidence that the US inquiry resulted in a conspiracy (by Bruce Ismay) to deviate it from the true sequence of events. The inquiry came close to unearthing the truth but ended up with a false series of events that prevail in the media and zeitgeist today. The United Kingdom inquiry was only set up to counterbalance the US inquiry in case it reached a different conclusion to one desired by the British government.

Reconstruct the Event Time-line

Having collected all the possible evidence as metrics, the US inquiry proceeded to reconstruct the event time-line right up to the point of impact and beyond. It was and still is standard naval practice to keep a running "ship's log" on board a ship, which is documentation of all sea and weather conditions, with all events carefully recorded, for all possibilities such as an inquiry.

Today, this is still the first step of a postmortem, namely, the discovery of what went wrong and what should have or should not have happened. For this step, you need to collect the evidence (metrics) and build the time-line of events.

Witness Testimonies

One challenge for the US inquiry was the ability to detain witnesses for a prolonged time, specifically, *Titanic*'s British crew, all White Star employees, who had to be subpoenaed, and who received no salary when on shore in New York. Therefore, the US inquiry had to get to the root cause quickly and gather information. This was a problem, as some information only revealed itself after several interviews. Other information seemed irrelevant at first but became essential once patterns emerged.

Figure 14.1. At the US inquiry, Lookout Frederick Fleet was a principal witness in providing details on the collision. He always maintained that had he been given a pair of binoculars, the ship might have been saved.[1]

In the US inquiry, the following expert testimony was identified:

"My theory would be that she was going along and touched one of those large spurs from an iceberg. There are spurs projecting out beneath the water, and they are very sharp and pointed."

—Extract from the testimony of Capt. James Henry Moore

Testimonies of Collision

Moore's testimony set up the myth that *Titanic* had a hard collision. In contrast, the following passenger and crew testimonies of the collision were identified:

"I heard this thump, then I could feel the boat quiver and could feel a sort of rumbling."

—Joseph Scarott, Seaman

"... It was like a heavy vibration. It was not a violent shock."

—Walter Brice, Able Bodied Seaman

"...I felt as though a heavy wave had struck our ship. She quivered under it somewhat."

—Major Arthur Peuchen, First-Class Passenger

"I was dreaming, and I woke up when I heard a slight crash. I paid no attention to it until the engines stop."

—C. E. Henry Stengel, First-Class Passenger

"We were thrown from the bench on which we were sitting. The shock was accompanied by a grinding noise...."

—Edward Dorking, Third-Class Passenger

"It was like thunder, the roar of thunder..."

—George Beauchamp, Fireman

A collision at 24 knots with a spur of ice would have been a "hard collision" and knocked everything sideways, resulting in many casualties with broken limbs. There is overwhelming testimonial evidence that this was a "soft collision," where the breakfast cutlery barely rattled.

The final report described the collision as:

> "The impact, while nonviolent enough to disturb the passengers or crew, or to rest the ship's progress, rolled the vessel slightly and tore the still plating above the turn of the bilge."

This implies a sideswipe of a spur, which has become one of the myths around *Titanic*. The US inquiry succeeded in identifying critical information, for example, the ship had reached a peak speed just before the collision, that Bruce Ismay interfered with the command of the ship, and that the basic feedback mechanisms were compromised.

US Inquiry Discrepancies

The US inquiry captured but failed to explain the following discrepancies:

- The testimony of Quartermaster Oliver indicated that Captain Smith rang down to the engine room "Ahead slow" ten minutes after the grounding, moving the ship forward.

- Senator Alden Smith discovered White Star had sent trains to Halifax, but these were canceled well after the ship had foundered.

Other evidence:

- Wireless radio operator Phillips sent a message to his parents twelve minutes after the grounding, "Making slowly for Halifax. Practically unsinkable. Don't worry." This was further evidence that the ship was moving after the grounding.

- The officers and key crew members provided testimonies that were very specific to the questions asked as though they had been coached.

Figure 14.2. At the US inquiry, wireless operator Harold Bride was another principal witness in providing details on the disaster.[2]

Figure 14.3. Senate Investigating Committee questioning individuals at the Waldorf Astoria. At the center is Carpathia wireless operator Harold Thomas Coffam.[3]

The cover-up over the grounding on the ice shelf and sailing off was not uncovered by the US inquiry. Although it collected the evidence, the inquiry failed to piece it together. Had it done so, the inquiry would have produced different conclusions.

Postmortem Step 1—The Discovery

Today, to complete step one (the discovery) of the postmortem, you need to create a problem statement and determine the factors. To better understand how to proceed with your postmortem, this section illustrates what should have been uncovered by the US inquiry into the *Titanic* disaster had the postmortem process been properly followed.

In a postmortem, once you have the time-line of events, you need to select the events that had the most impact on the disaster—those thought to be "problematic"—before you can start to discover root causes.

Problem Statement

Selecting these key events requires careful definition of a problem statement, as there might be ambiguity around which problems are important, or if several problems occurred simultaneously.

For example, for a postmortem about *Titanic*'s disaster, a first question about the disaster might have been "Why did *Titanic* not reach New York?" The answer is that it ran onto an ice shelf and foundered after it sailed off. However, it is important to phrase the question in a meaningful context to the organization and, in this case, the tragic and huge loss of life, so the question should have read, "Why was *Titanic* such a tragic disaster, with so many lives lost?" The emphasis should have been on learning lessons and preventing future catastrophes. After all, the public had relatively little sympathy for White Star and its economic loss.

Factors

A good problem statement helps determine the factors in *Titanic's* disaster:

- Officers failed to slow the ship and prevent the collision with the ice shelf.

- Bruce Ismay and Captain Smith sailed the crippled ship off the ice shelf for Halifax (based on the above discrepancies).

- Both officers and crew undertook recovery poorly.

Postmortem Step 2—The Analysis

In your postmortem, you are now ready to proceed with an analysis (step two) that will lead you to the root causes of the disaster. This step determines why things went wrong. The first element of this step is to select the events that most affected the disaster, or were most "problematic" related to the factors. For example, for the first factor (officers failed to slow ship), these were the following:

Selecting the Critical Events for the First Factor

These were (for officers failed to slow ship) the following:

- Seven ice warnings were received during the day but were only passed on an ad hoc basis to the bridge.

- The last ice warning from *Mesaba* was not passed to the bridge.

- Visible signs of ice around the ship were ignored.

- Lookouts spotted ice looming 50 feet above the water, but could not prevent a collision.

Selecting the Critical Events for the Second Factor

These were (for crippled ship was sailed off ice shelf):

- Water damage in the fore-peak and cargo hold was not taken seriously enough.

- Bruce Ismay arrived on deck to check damage and assumed authority over Captain Smith.

- The first investigation group led by Officer Boxhall returned with an inaccurate damage assessment.

- Bruce Ismay and Captain Smith restarted the engines acting on sparse information, reluctant to wait for the report of the second investigation group.

Selecting the Critical Events for the Third Factor

These were (for recovery undertaken poorly by officers and crew):

- Executives knew that the ship had few hours left, but that information was not widely communicated.

- No "abandon ship" order was given; the information flow was very slow or even restricted.

- The first distress call was sent out 40 minutes into the disaster, showing a reluctance to face the situation.

- The first distress rocket was sent 65 minutes into the disaster, confirming a reluctance to face the situation.

- The first few lifeboats left half-empty 65 minutes into the disaster, and the crew could not persuade passengers to board them.

Analysis of the First Factor

This is the next step in the postmortem process. This showed (officers failed to slow ship):

- A lower speed would have prevented the disaster, an opinion shared by naval architects and engineers. The iceberg's impact would have been minimized, and with less damage to the hull plates, fewer compartments would have flooded.

- The officers were so confident in their ship's handling that, despite many warnings, they chose to press full speed ahead, possibly because:

  - Of the experience gained with *Olympic,* which had been in operation for 11 months, similar conditions had been encountered.

  - The built-in feedback mechanisms would provide early warning signs of ice.

  - Many ships sailing along busy sea routes would provide ample radio message warnings of ice.

  - The visibility conditions were perceived as excellent.

The sea was very still, and the night sky very clear, so danger should have been seen in advance.

Analysis of the Second Factor

This (crippled ship was sailed off ice shelf) showed some interesting causes:

- Bruce Ismay and Captain Smith, convinced that no serious damage had been sustained, restarted the engines and floated the ship off the ice shelf, hoping to limp back to Halifax. They chose to do so, possibly because:

    ○ The collision seemed so innocuous, with no signs of visible damage and no casualties. The impact was no more than a vibration and grinding noise that grounded the ship.

    ○ The resounding belief was that the safety features could cope with the damage. The water pumps and double hull kept up with the flooding, and the bulkheads were sealed.

    ○ The first damage assessment report was incorrect because Officer Boxhall's group did not descend enough decks to observe the damage properly.

    ○ Bruce Ismay and Captain Smith were too impatient for the second damage assessment report because of business pressure to arrive in Halifax within a specific period to the estimated arrival time. Bruce Ismay saw an opportunity to extend the legend of Olympic-class ships further, in having the ship save itself. He was under business pressure to maintain the perception that *Titanic* was the greatest liner and to avoid embarrassing publicity if distress calls were put out stating that the ship was grounded on an ice shelf.

Analysis of the Third Factor

This (recovery undertaken poorly by officers and crew) showed the following items:

- The officers and crew could not enact a smooth recovery. The statistics for third-class passengers lost were horrendous compared with the total aboard.

- Had the officers and crew filled the lifeboats properly, an additional 750 people could have been saved. They did not, possibly because:

    - No plan (similar to business continuity) was in place to provide some guidelines for the recovery. A plan was thought unnecessary because of the confidence in the ship's safety features.

    - The crew was preoccupied in helping first-class passengers to the boat deck and lifeboats.

    - The cascading information flow was very slow, and Captain Smith failed to give concrete orders such as "abandon ship." Sometimes, the crew lied to the third-class passengers about the extent of the damage to encourage a return to their quarters.

    - Officers and crew responded slowly because of the disbelief that the ship was in danger, and they only realized something was wrong as disaster signs appeared after the first hour. There was overconfidence in the ship and its safety features.

- Third-class passengers had great difficulty in getting to the boat decks because:

    - The class system on the ship exasperated the problem of movement. The lower the class, the farther and deeper into the ship was the accommodation.

    - Third-class passengers were physically restricted by gates and barriers from wandering outside their class because US immigration regulations specified that immigrants on ships be segregated from other classes for reasons of health.

- ◦ Many of these gates were locked while the ship sank because the crew members in charge of the gates were unavailable to open them.

- ◦ Many third-class passengers themselves accepted their "lower position" in the socially created hierarchy, which further reinforced the segregation.

Postmortem Step 3—Corrective Actions

In your postmortem, you are now ready to proceed with the third step that tests and rationalizes the root causes into true causes and determine solutions or corrective actions. The purpose of this step is to prevent the disaster from happening again.

Causes for the First Factor

For the first factor (officers failed to slow ship), the root causes were the following:

1. Clear procedures were lacking, for example, procedures for passing ice warning messages to the bridge and for the officers to chart the ice field's size. Had one of the officers been able to put together all the ice warnings, it is very likely the size of the ice field would have been better understood.

2. Business pressure to better *Olympic*'s crossing time influenced action. Bruce Ismay was determined to prove that *Titanic* was a better ship. Sea conditions were calm and so were treated cavalierly, whereas other ships in the region had pulled up for the night.

3. Data from feedback mechanisms was distrusted, discounted, or just ignored by the officers if it didn't fit their perceptions. Some data, such as the result of the ice detection test, were fabricated. The captain, a technophobe, tended to operate by instinct.

4. Officers saw no reason to post more lookouts to the ship's bow, as visibility was perceived to be excellent. In reality, it was very poor because of the haze caused by the cold weather, and with the calm sea, there were no breakers to identify icebergs on the horizon. Lookouts were posted without binoculars, while officers kept theirs.

Root Causes for the Second Factor

The second root cause (above) is likely the true cause of the disaster's first factor. A remedy could have been to play down the significance of the maiden voyage and have a "burn-in period" of shorter trips. Then, later in the summer, when icebergs were less likely, reschedule the voyage and take a more southerly route.

However, there were great financial pressures to get *Titanic* operational quickly. For the second factor (crippled ship was sailed off ice shelf), the root causes were the following:

1. Officers and crew disbelieved the seriousness of the situation. Even after an hour, there was still disbelief in the seriousness of the situation by officers and crew, a common disaster trait.

2. There were strong business pressures to maintain the perception that *Titanic* was the greatest liner. Bruce Ismay callously exploited a very risky opportunity to save face by turning a problematic situation into a positive one, in his mind.

3. The first damage assessment group was not expecting any serious damage. The investigation was hurried and not done thoroughly. The executives looked for exactly this type of evidence to use to their advantage.

4. Executives could override procedures and the rules of good seamanship. Bruce Ismay used his position to control the operation and all key operational decisions, without adequate relative experience, to the detriment

of the organization. Bruce Ismay intimidated Captain Smith, even with the overwhelming evidence of the dangers. The overriding authority of Bruce Ismay, Captain Smith, and the hierarchical culture made it impossible to challenge.

Root causes 2 and 4 are likely the true causes of the second factor. A remedy could have been to establish SLAs (to deliver passengers safely) and then establish responsibility firmly with one group alone (operations services) to meet these without interference from executives whatsoever.

In a crisis, it is easy to lose control and make rash decisions on too little information. Hence, the group responsible for operations services needs to follow processes such as a problem management process. They would have likely refused to act before examining more evidence, and the ship could have stabilized on the ice shelf, which would have been long enough for rescue ships to complete a full recovery of most of the passengers.

Root Causes for the Third Factor

For the third factor (recovery undertaken poorly by officers and crew), the root causes were the following:

1. A fear of widespread panic. This was inevitable, as many third-class passengers on the lower decks, where the flooding was most apparent, were well aware of the disaster very early. In many organizations, customers first notice outages.

2. Communication systems were very poor. The crew had to alert passengers by knocking from door to door, which took up to an hour. On the upper decks, a level of calm was kept, as the crew was close by and helpful, unlike the lower decks, where the signs of disaster were more visible.

3. No business continuity plan was in place. A disaster recovery plan had not been carefully thought out for all scenarios.

The third root Cause is likely the true cause of the third factor. A remedy could have been to establish a disaster recovery plan, carefully thought out for all possible scenarios. The plan would be well communicated and regularly practiced.

What You Can Learn from Root Causes

In reviewing *Titanic*'s postmortem, eleven significant root causes were identified. Nine of these were related to organizational issues, two were related to the lack of procedures or plans, and none was related to technology.

The following four root causes were most significant:

- The business pressures to better *Olympic*'s crossing time influenced actions, as Bruce Ismay was determined to prove *Titanic* was a more technically advanced ship.

- Bruce Ismay used his authority to control the operation and all major decisions, even though stringent guidelines and procedures were in place.

- No business continuity or disaster recovery plan was in place, as it was thought unnecessary.

- The business pressure to maintain the perception of the greatest liner caused Bruce Ismay to exploit a very risky opportunity callously to save face.

Clearly, the true causes were related to business pressures overriding operational decisions and very poor leadership. This can occur in today's operations, where business pressures and overzealous leadership can override operational decisions. You should complete postmortems for all projects and major problems, closely look at the root causes, and not accept superficial excuses.

In today's popular culture, we have come to accept that *Titanic* was just unfortunate. The truth behind the root causes was falsified for political reasons. The failure of White Star's captain, officers, and crew to protect the ship should have resulted in a case of gross incompetence being brought against White Star. In effect, the British investigation knew that it would have to exonerate White Star to stave off bankruptcy and prevent rival German companies from dominating transatlantic shipping. Most important, White Star ships were essential in time of war for transportation.

US Inquiry and Changes to Maritime Transportation

A disaster of *Titanic's* proportion was probably the only single event that could shake up shipping companies into making changes and introducing safety standards to keep up with changes in emerging technology. The US inquiry led to the following changes:

- The International Conference for the Safety of Life at Sea (SOLAS) approved this following resolution in November 1913: "When ice is reported on or near his course, the Master of every vessel is bound to proceed at night at a moderate speed or to alter his course, so as to go well clear of the danger zone."

- In 1913, the International Ice Patrol organization was created, financed by the nations that used the North Atlantic shipping lanes. The sea lanes were patrolled during greatest iceberg danger, in the January-to-August period.

- The number of lifeboats was increased on ships to "a place for every soul."

- After the sinking, Morse code was installed as the standard communication for ships at sea by most maritime nations. At an international conference, convened three months after the disaster, the SOS distress signal was adopted. The signal was adopted because it was an easily recognizable letter sequence of three dots, three dashes, and three dots. However, it became popularly known as "save our souls" or "save our ship."

- The wireless became accepted as an important safety device. Each ship required two radio operators to cover the 24-hour day. No longer was the wireless seen as a tool to entertain passengers.

- The southern Atlantic route was moved even farther south by 60 miles during the summer, to avoid any risk of iceberg collision.

- The German liner *Imperator*, built in 1912 after the disaster, was delayed by the addition of an inner skin, which extended above the waterline in the forward compartments. The space between was five feet and was filled with water to test the tightness of the skins.

The British Inquiry a Whitewash

Officer Lightoller, in his biography, talked about holding a white brush. He was referring to the British investigation as a whitewash and his collusion. It shifted the blame to the Board of Trade for not changing the lifeboat regulations to keep pace with increases in ship size and the captain of *Californian*, Captain Stanley Lord, who sat the night out. He was surrounded by ice, unaware of the disaster, and he did not come to *Titanic*'s rescue. However, even if *Californian* had come to the call, it is highly likely that the rescue attempt would have been unsuccessful. Why did the British government need White Star to stay in business? It saw a potential European war looming (1914–1918), and knew it would need large ships for transporting troops and materials. Today, *Titanic's* disaster would undoubtedly have brought White Star down just through private lawsuits. In the business world, this has repeatedly happened. You need to consider these implications and their impact carefully when implementing the solution into production.

Figure 14.4: Surviving officers Pitman and Lightoller in New York City during the US Inquiry 1912.[4]

Post Inquiries

The British Government saved White Star. To survive as a business, White Star had to act quickly and make changes to reassure the public and customer base.

Lawsuits

In the United Kingdom, the British inquiry judge did not find White Star at fault. He exonerated Bruce Ismay, Captain Smith, and the officers. Lawsuits could only be filed in the US. Senator Alden Smith failed to pin White Star responsible for the disaster. In the US, private lawsuits were pressed. The resulting claims were for $17m ($357m today). White Star settled the claims out of court but paid a fraction of the suits—about $660,000 in USD ($14m today).

Retrofitting the Ships

Special Cable to the Examiner.

> Belfast, Sept. 24—The White Star Line announces definitely that the steamer *Olympic,* sister of the *Titanic,* will come to Belfast from Southampton for renovation. She will be made identical with the new ship *Britannic,* now building. The *Olympic* is due at Southampton October 5.
>
> —*Chicago Examiner,* Wednesday, September 25, 1912,
> p. 2, c. 2.

Both *Olympic* and *Britannic* were fitted with an extra bulkhead to take it up to 17 watertight compartments. Five of the bulkheads were extended 40 feet above the waterline up to bridge deck.

Figure 14.5. Retrofit of the boat deck saw an increase in the number of lifeboats.[5]

Figure 14.6. Britannic went through a retrofit, with a substantial increase in lifeboats; bulkheads were altered running fully from top to bottom.[6]

Fate of White Star Line

Titanic's sinking in 1912 was a financial disaster for IMM. The company labored on but had to apply for bankruptcy protection in 1915. A poor cash flow caused IMM to default on bond interest payments, and with $3.3 million in interest owed, IMM was declared to be in technical bankruptcy where a receiver was appointed. Mr. Franklin succeeded to the presidency of IMM. Only World War I saved IMM because of wartime demands for shipping. IMM steamers carried one-quarter of the entire American Expeditionary Force to France and nearly 15 million tons of war supplies.[7]

What If Scenarios

Californian Rescue Scenario

Californian's role has been at the heart of a controversy, and Captain Lord was heavily criticized in both the inquiries. The main question today is whether he could have mounted a rescue. *Californian* was a few hours away but had pulled up for the night, as it was too dangerous to sail through ice-strewn waters. Earlier, Captain Lord had avoided a major collision with ice by ordering full speed astern to stop on the edge of an extensive ice field.

Captain Lord had been proactive and had instructed his radio operator Evans to send ice warnings to *Titanic*. Unfortunately, Evans was snubbed, so he turned off his radio wireless and went to bed.

Californian was estimated to be between 8 and 19 miles from *Titanic*. At 12:45 a.m., the crew saw *Titanic*'s rockets, but these were misunderstood. Captain Lord was informed at 1:10 a.m., but he concluded the ship had stopped for the night, and all were having a party.

What if Captain Lord had responded differently? What would a rescue scenario with *Californian* have looked like? *Titanic* took about 2 hours 40 minutes to sink (11:40 p.m.–2:20 a.m.). Captain Lord had more than one hour to mount a rescue. *Californian*'s best speed was 13 knots, so the earliest arrival time would have been about 2:45 a.m., with consideration for clearing the ice field.

DISTANCE Miles	SPEED	DURATION	TIME TO CLEAR ICE	ESTIMATED ARRIVAL TIME
19	13	1:24	0:30	3:05 a.m.
8	13	0:59	0:30	2:40 a.m.

Figure 14.7. Estimate of the possible arrival times based on distances between the two ships indicated that, even at the shortest distance, it would have been challenging for Californian to arrive in time to make a difference. The time does not account for Californian's launching her lifeboats.

Californian would have arrived to face a scene of 1,500 people in ice-cold waters. Transfer at sea was time- and labor-intensive, and pulling up against *Titanic* and transferring people across was unfeasible. Only a lifeboat rescue could be completed, but using her 6 lifeboats as ferries required a crew of 48 (8 x 6 lifeboats). She had only 29 seamen, officers, cooks, and stewards on board. *Californian* was a very small ship (6,233 GRT) and could carry, at tops, 218. At best, Captain Lord could launch only two lifeboats, with 12-man crews. With two lifeboats, there was the problem of 1,500 survivors trying to get 1 of 120 lifeboat seats available. The 12 men in each lifeboat would have to fend off 1,500, and they would have been swamped, or survivors trying to board both sides of lifeboats would have likely overturned them. The rescue would most likely have failed.

Besides, Captain Lord's primary responsibilities were to his ship. He did not automatically have to rush to *Titanic's* aid that fatal night, and being risk-averse, he chose to pull up for the night.

Chapter Wrap-up

Conclusion

One of the most significant facts missed by the US inquiry was that *Titanic* was passed off as a British ship, which prevented liability held against Bruce Ismay. It is difficult to believe Senator Alden Smith did not associate White Star with J. P. Morgan and IMM. The question is whether Senator Smith skirted the issue that White Star was under the umbrella of a US company. Surely, he would have known the connection.

In the US inquiry, Senator Alden Smith and his team were thrown off the trail by "mariner speak," nautical terms used by officers and crew in their testimonies. Reading over these today leaves the impression that certain important points were not fully understood by the inquiry that failed to pursue and investigate these further, as they should have been.

Key Lessons for Today

Today, a project manager should do the following:

- Confirm that a project postmortem is supported by the organization and endorsed by the PMO. This should be done in the initial stages of a project.

- Ensure the postmortem gets at the truth, and it is not held back politically. This requires governance to be in place to provide the postmortem with the right stature and authority. This should be checked early in the project.

- Confirm that time is set aside to go through a postmortem once the project is completed.

Educators

Discussion points:

- Was Senator Alden Smith the appropriate person to lead the US inquiry?

- Was Captain Lord morally justified in not coming to *Titanic*'s rescue?

- At what range of distance could *Californian* have completed a successful rescue?

- Are there are both criminal and commercial liability repercussions on the US inquiry's failure to realize that *Titanic* wasn't British owned?

- In reviewing *Titanic*'s postmortem, the eleven significant root causes were identified. Nine of these were related to organizational issues; two were related to the lack of procedures or plans; and none was related to technology.

CHAPTER 15

Conclusion

Conclusion to the Book

The most striking thing about the *Titanic* case study is how the roots of the disaster can be attributed back to the project's first day. The project team's compromises, in the design and construction phases, had an impact later in the project that set the seeds for failure. These compromises were preventable, but because they could pass, they became difficult to reverse in later phases of the project.

As the project moved into implementation and operation, further compromises were made, and the project team bowed to business pressures to get the ship back into operation quickly. However, a complacent attitude had evolved in the project, believing that even if things went wrong operationally, it did not matter. The view prevailed that the ship was so technically advanced that it would be protected against any worst-case scenario encountered, creating a project atmosphere where making compromises and taking risks became more acceptable.

The main conclusion of the book is that White Star was not unfortunate or incompetent with this project. A continuous sequence of compromises through the project created a situation where a disaster was inevitable. Unchecked egos, uncontrolled agendas, fear of disagreeing, and the inability to step in and manage the project made this project one of the most infamous failures of the twentieth century—a major lesson for all project managers today.

Recap and Key Findings

The project to build the Olympic-class ships was begun on sound footing. White Star embarked on a well-thought-out strategy to invest in emerging technologies and create three superliners. White Star's relationship with Harland and Wolff was critical in this project, and White Star saw no need to tender the contract.

From the outset, the project scope was not seen as daunting, but achievable, based on the experiences of White Star and Harland and Wolff, who had made massive investments in upgrading its shipyard facilities.

As the design neared completion, the project architects elected to go with the highest level of safety and incorporated the latest safety technologies, including a double hull, bulkheads with electric doors, and triple-stacked lifeboats. But the architects were undermined by executive pressure from Bruce Ismay, who pushed for the ultimate passenger experience and a spacious ballroom that cut straight across several bulkheads. Decisions were made that compromised the initial vision and the safety features. Most disturbingly, the architects allowed these compromises to pass.

As the project went through the construction phase, the project team still believed the Olympic-class ships were practically unsinkable and could survive any worst-case scenario because of the aggregated effect of safety features, the broad hull design, sheer size, and the latest technologies. This was also used actively in the marketing campaign. The lifeboats were viewed as an extra safety feature to be used to ferry passengers from other ships in distress.

In parallel, Bruce Ismay's marketing plan took effect and at *Olympic*'s launch, huge crowds were attracted, even though the ship was a year from completion. The Olympic-class was marketed to the public as practically unsinkable, which became widely accepted.

Olympic's career was marked by several incidents. When *Olympic* was very much damaged through the *Hawke* incident, Harland and Wolff came to the rescue and got her back into service in record time, saving the reputation of White Star that, by now, was completely dependent on Harland and Wolff. The incident was a huge cost to White Star, and it compromised *Titanic*'s project phase, the fitting-out. Harland and Wolff's repairs highlighted how deep and important the relationship was between the two companies. Any questions over the decision not to tender the contract were answered by how Harland and Wolff responded and went beyond expectations as a loyal and true partner to White Star.

Harland and Wolff had performed near miracles in repairing *Olympic* with complex and extensive repair work that went beyond the project's scope. Harland and Wolff also pulled off the near impossible and delivered *Titanic* only one day late. The ship underwent one day of sea trials, and with the established perception that the ships were invincible, *Titanic* was rushed into service. In reality, *Titanic*'s testing consisted of the maiden voyage across the Atlantic, fully loaded with passengers. The contract was not properly fulfilled, but White Star was very much driven by the pressing economic need to have two ships in service.

With the phased delivery of three ships, Bruce Ismay saw a marketing opportunity to promote each ship as an improvement over the last. Bruce Ismay's publicity stunt to beat *Olympic*'s best crossing would allow him to market *Titanic* as superior. But this was a further compromise, as Bruce Ismay wrote out a new service-level goal without verifying it with his captain and officers, which was fateful in pushing the ship to her operational limits.

On leaving Southampton, *Titanic* had a near collision, similar to the *Olympic/Hawke* incident. The steamer *New York* came within four feet of striking *Titanic*, indicating the challenges in handling the large ship. When *Titanic* got to Queenstown, the poorly executed lifeboat drill before the

Board of Trade inspectors failed to alert anyone that the crew was unprepared for a disaster that would require the launch of all lifeboats.

At sea, the officers and crew were still acclimatizing to the ship and completing some testing missed in the sea trials. They also had to deliver a maiden voyage that would inspire the world's elite on board. But Bruce Ismay's marketing effort, centered on the superiority of the Olympic-class ships, was almost too successful. It instilled a mind-set in the officers, crew, and passengers that even if things went wrong operationally, the ship had enough safety features to protect it in any scenario. In other words, the ship was practically unsinkable.

The voyage was riddled with problems caused by compromises in the implementation and operation phases. The operational feedback mechanisms did not work properly. Compromises in the ice detection test and with the lookouts were serious. More alarmingly, the radio operators only sporadically relayed warning messages to the bridge because of the flood of outgoing commercial radio messages. There was a serious oversight in how the radio operators reported into the ship's hierarchy. In addition, Bruce Ismay patrolled the ship with a cavalier and arrogant attitude, ignoring operational procedures, and pushing the crew to reach maximum speed. This violated all the basic rules of good seamanship. Class boundaries existed between the crew and officers, and the captain and director. Officer Murdoch was under tremendous pressure not to slow, and he was forced to navigate a very risky run through a hazardous ice field. It was a disaster waiting to happen.

The collision was almost inevitable. Officer Murdoch came close to preventing it through brilliant seamanship, when he almost pulled off an S-turn. The ship's operation was pushed beyond the limits for which its safety features were designed. The collision was so innocuous that it was not taken as seriously as it should, leading to the most serious of compromises, the decision to restart *Titanic* and move her off the ice. Bruce Ismay's anxiety over White Star's reputation made him hellbent on saving face and what greater feat than *Titanic* saving herself. Further mistakes such as delaying the launch of the lifeboats and not communicating the reality of the situation to passengers led to six of the first eight lifeboats leaving half-empty, which compounded the disaster into a horror.

At the two inquiries, Bruce Ismay and the remaining officers supported the ice spur story to hide the truth. This had in fact been put forward by one of the US expert witnesses. The British government helped the cover-up and saved White Star from bankruptcy. After all, with the Great War looming, Britain needed large ships for transportation.

Learnings for Today

What can we learn and take forward from this analysis. One of the most significant learnings is all the compromises and mistakes were preventable. Early in any project, in the initiation and planning phases, serious thought needs to be given to the operation of a solution, with a walkthrough of possible scenarios that could affect this and cause a disaster.

Another learning is not getting too close to your supplier. White Star's relationship with Harland and Wolff should be questioned. If the two organizations had not been so interdependent then Harland and Wolff would have been more likely to say no on occasions. A project needs to stay true to its vision and direction through each phase of the life-cycle. It is unfeasible for a project manager to micromanage the whole project. As a result, the project team must be empowered to do this. It has to understand the project vision clearly, aggregate information, determine risks, and present critical issues at steering committee meetings. Strategies such as starting with a pilot and then scaling each project life-cycle rapidly minimize the impact of poor decisions. Reviewing the business case through the project is important as is looking at the bigger economic picture around the business case. White Star failed to consider the period of the long construction project of six years and the possible changes that could affect the liner business, through events or technology.

Confidence that nothing can go wrong can create an at-mosphere where little attention is paid to the operational readiness. This might account for the persistently high project failure rate (25% to 30%) still seen, a figure that has been continuously verified in various surveys.[1] The success of projects today should not be measured at deployment, but rather after the solution has been in production for a while and carefully measured. Metrics should be closely tied to the overall impact to the business. The *Titanic* story helps us better understand the relationship between functional and nonfunctional requirements, the interplay of compromises in the project, and why things go very wrong in implementation and operation.

Titanic's case study has shown that root causes to many operational problems originated in the project. Bring this forward to today, and many comparatives can be made to modern projects, from construction right through to production. For example, there are many similarities with how project problems and issues surface days, months, or even years after the project is completed and in production. This was my own experience with a project postmortem in which I was involved in the mid-'90s and started me on the path of investigating the *Titanic* case study.

Project managers today need to complete diligently each project phase from planning to constructing to testing to implementing and operating. It is easy to dismiss compromises in these phases and accept the deliverable as long as it can be signed off. On implementing the solution successfully, it is easy to become very complacent, assume nothing will go wrong, declare the project finished and a success, and move on to the next project. So when should a project shut down? When can a project manager walk away? The most common answer is when the deliverables are signed off, but the real world is not this simple, and project managers must deal with gray situations.

Final Statement

The most important thing to gain from this book is that project management is about people and communication—understanding stakeholders and their expectations, and then managing these, so they stay aligned to the direction.

Profiles of characters

Joseph Bruce Ismay (1862—1937)

Bruce Ismay was the oldest son of Thomas Henry Bruce Ismay, founder of the White Star Line. After completing his education, he served an apprenticeship at Thomas Bruce Ismay's office for four years. He was posted to the New York White Star Line office where he was eventually appointed the company agent in New York. He married, and when he returned to England, he was made a partner in the family firm. When his father died in 1899, he became head of the business. In 1901, he negotiated terms with J. P. Morgan, through which White Star Line became part of the International Mercantile Marine Company (IMM), an international conglomerate of shipping companies. In 1904, he became president of IMM. Aside from this, during his life, he became chairman and director of many companies and associations in shipping, insurance, canals, and railways.

Figure A.1. Joseph Bruce Ismay.[1]

Bruce Ismay started the project to build the Olympic-class ships. As the owner of White Star Line, he had a critical part in the whole project, and he was the most important player in the overall project. He was the project sponsor responsible for beginning the project, articulating the vision with Pirrie, and building the business rational/case for the venture. He took a keen interest in the project, and he was the principal stakeholder, representing the White Star board and responsible for signing off on deliverables.

Bruce Ismay had a very specific plan related to the vision he helped create, which centered on building a reputation for the three liners as the end word in luxury. In targeting the first-class passengers, he shaped everything around the ultimate passenger experience. However, this proved damaging overall because he interfered with decision processes through the design and construction of the project. In reality, his meddling forced compromises to be made beyond what could be recognized as reasonable. Further, his drive in securing publicity around the maiden voyage led to risks taken to prove *Titanic* was technically superior to *Olympic*. Bruce Ismay adopted the role of pseudo-captain on the maiden voyage, pushing the crew to hit maximum speed through the treacherous ice-strewn waters.

Post disaster, the papers called Bruce Ismay "The Most-Talked-of Man in the World" chiefly because he saved himself. This was a dubious honor when the title was put in conjunction with terms as "public opinion," "on trial," and "pariah." Bruce Ismay was savaged by the press, both in the

US and the United Kingdom for deserting the ship while women and children were still on board. When he returned to England, he was shunned by society, as his reputation was irreparably damaged. He resigned his position in 1913 when Harold Sanderson took over and maintained a low profile. He retired from active affairs in the mid-1920s and retired to County Galway, Ireland. He died on October 17, 1937 leaving an estate worth £693,305.

The *Times* obituary recalled Bruce Ismay's personality:

[He was a man] of striking personality and in any company arrested attention and dominated the scene. Those who knew him slightly found his personality overpowering and in consequence imagined him too be hard, but his friends knew this was but the outward veneer of a shy and highly sensitive nature, beneath which was hidden a depth of affection and understanding which is given to but few. Perhaps his outstanding characteristic was his deep feeling and sympathy for the 'underdog' and he was always anxious to help anyone in trouble. Another notable trait was an intense dislike of publicity, which he would go to great lengths to avoid. In his youth, he won many prizes in lawn-tennis tournaments; he also played association football, having a natural aptitude for games. He enjoyed shooting and fishing and became a first class shot and an expert fisherman. Perhaps the latter was his favourite sport and he spent many happy holidays fishing in Connemara.

William James Pirrie (1847–1924)

Pirrie was a Canadian-born British lord. He was chairman of Harland and Wolff but also a director of White Star. He (a working-class boy) started with the shipbuilding firm as an office boy and worked his way through the company as an apprentice, so he knew all aspects of the business. He worked his way up to the top, gaining a reputation as a hands-on boss who oversaw every step of construction and frequently made unannounced inspections of projects. He was chairman of Harland and Wolff between 1895 and 1924 and served as Lord Mayor of Belfast between 1896 and 1898.

Figure A.2. William James Pirrie.[2]

As the Chairman of Harland and Wolff, the shipbuilders, Pirrie worked with Bruce Ismay to begin the project to build three superliners back in 1907. Pirrie, over dinner with Bruce Ismay, articulated Ismay's vision for three superliners sweeping the Atlantic. Both men recognized that White Star's current fleet of liners was hopelessly outclassed by the competition Cunard Liners, which was a class of ship built for speed and for capturing the prestigious Blue Riband. Pirrie, the project integrator, helped shape the brilliant strategy of providing a clear differentiator of luxury or comfort over speed based on using the latest in emerging technologies. Pirrie remained involved in the project until 1909. Illness prevented him from traveling aboard *Titanic* on her maiden voyage.

John Pierpont (J. P.) Morgan (1837–1913)

Born to a father, who was a prosperous international banker with a home on Asylum Street in Hartford, and a mother with roots in Connecticut Puritanism, young Morgan was educated at a school in Vevey, Switzerland, and at the University of Gottingen. Morgan went into the family business in London in 1856. He started his own private banking company in 1871, which later became known as J. P. Morgan & Co. in 1895. He became one of the most famous financiers in the history of business and nothing less than the nation's financial backstop. When the gold reserves of the United States became dangerously low in the mid-1890s, President Cleveland turned to Morgan to oversee the international purchase of $65 million in gold. Morgan handled the operation to the advantage of the financial stability of the nation and to the enormous profit of J.P. Morgan and Company.

Figure A.3. John Pierpont (J. P.) Morgan.[3]

Physically imposing with a large frame, massive shoulders, and piercing eyes, Morgan could easily intimidate by his presence alone. That presence combined with a frequently abrupt and dictatorial bearing made him a formidable competitor—even at dinner parties.

The flow of investment in the United States was thus directed to, and the expansion of industrial capacity took place in, industries and firms Morgan and his few peers wished to see expand, not elsewhere. Morgan and his partners built the mold in which American turn-of-the-century industrial development was formed. The New York Central, Northern Pacific, Erie, and AT & SF railroads issued their bonds under Morgan auspices and had Morgan representatives on their boards of directors. Morgan partners had strong voices in the selection of management for, and in the choice of corporate strategy of, AT&T, General Electric, and Westinghouse. Morgan masterminded the merger that created US Steel in 1901. He gathered the individual railroads of the United States into continent-spanning systems.[4]

He was criticized for creating monopolies and making it difficult for other businesses to compete with his. Morgan dominated both the railroad industry in the East, which he helped consolidate, and the United States Steel Corporation, the world's largest steel manufacturer. The government, concerned that he had created a monopoly in the steel industry, filed a suit against the company.

Morgan was used to good living, large yachts, and three-month tours of Europe: "I can do a year's work in nine months," he would say, "but not in twelve." This could have been a driving factor in acquiring a conglomerate of shipping companies to put under International Mercantile Marine (IMM). He was in the transportation business, and with his railroad empire, moving into shipping was a natural progression.

He was widely criticized for his financial dominance in the Panic of 1907 and for bringing the financial ills of the New York, New Haven & Hartford railroad. As the panic approached, Morgan met with his partners, financiers, and industrialists. He directed the banking coalition into stopping the Panic of 1907. He was unconcerned about popular criticism.

In 1907, he was arguably the richest man in the world. As chairman of IMM, he owned a controlling interest in White Star Line. The public believed White Star was a British shipping line because of the colors flown and the British officers and crew employed, but *Titanic* was US-owned and part of Morgan's strategy to leverage the best in global technology. Although not immediately identified with *Titanic's* story, Morgan provided the capital for the three superliners to be built. His connections with the noveau riche class in the United States also helped draw the who's who list of millionaires for the maiden voyage. Morgan was due to sail on the maiden voyage himself but canceled the night before because of illness. Morgan can be viewed as the project financier and principal executive.

Post disaster, he was not implicated in the project, and he distanced himself from it. In 1912, he appeared and publicly defended himself before a congressional committee, which was investigating the money-trust particularly aimed at him.

He died in Rome in 1913, at the age of 75, and left his son, John "Jack" Pierpont Morgan Jr. to run J. P. Morgan &

Co. The New York Stock Exchange was closed until noon, the morning of his funeral, an honor reserved for heads of state.

Read more: Morgan—Infoplease.com http://www. infoplease.com/ce6/people/A0834017.html#ixzz1ZUXQPCpV

Thomas Andrews (1873–1912)

Andrews, a native of Northern Ireland, was a naval architect at Harland and Wolff. He was also the nephew of Lord Pirrie who decided not to show any favors whatever based on the relationship. Andrews, through his own efforts and abilities, had to make his way, and if he failed, that was as many others who had gone before him. He served a five-year apprenticeship and went through the various departments learning rigging, plating, and engine building. He spent three months in the joiner's shop, one month with the cabinetmakers, two months working in ships, two months in the main store, five with the shipwrights, two in the Molding Loft, two with the painters, eight with the iron shipwrights, six with the fitters, three with the pattern makers, eight with the smiths, followed by eighteen months in the drawing office, which completed his term of five years as an apprentice. He was motivated enough to take additional classes in mechanics, engineering, and marine architecture in his own time after twelve-hour workdays. With energy and enthusiasm, he rose through the ranks of Harland and Wolff to become the construction manager and head of the design department.

Andrews played a significant role in the Olympic-class ship project, and he was accountable for turning the vision into reality. Involved from the outset of the project, he took a lead through the four-year design and construction of the project. He also personally designed the complex inner steel framework. A diligent man, he carefully planned for all aspects of the operation and a recently discovered personal notebook highlights his foresight as he planned for 48 lifeboats, a place for everybody on board.

Figure A.4. Thomas Andrews.[5]

He was on both *Olympic*'s and *Titanic*'s maiden voyages. With *Titanic*, he represented Harland and Wolff in place of Lord Pirrie, who was ill at the time. He was always seen with a notebook carefully noting flaws that would be rectified before the next transatlantic crossing.

He busied himself capturing the smallest of details. Shan Bullock writes: "For more than a week, he had been working at such pressure, that by the Friday evening [April 12, 1912], many saw how tired as well as sad he looked: but by the Sunday evening, when his ship was as perfect, so he said, as brains could make her, he was himself again." Later that evening, *Titanic*'s collision with ice sealed the ship's fate and that of Andrews and two-thirds of the people on board.

Andrews was the first person the ship's captain turned to for advice after the ship's collision. He quickly determined that the ship was sinking, after inspecting the damage. Andrews told Captain Smith there was little to prevent it.

Andrews acted heroically, helping people to the lifeboats. Bullock records a popular legend, writing that, after 2 a.m., "An assistant steward saw him standing alone in the smoking room, his arms folded over his breast and the [life] belt lying on a table near him. The steward asked him: 'Aren't you going to have a try for it, Mr. Andrews?' He never answered or moved, 'just stood like one stunned.'" There is no evidence that he tried to save himself.

Alexander M. Carlisle (1854–1926)

He was the shipyard's general manager and chief draftsman when the *Olympic* and *Titanic* were ordered. He was responsible for coordinating the designs, and his main area was the equipment used on the ships. Carlisle proposed a new type of davit developed, capable of holding four wooden lifeboats, to give on board enough to handle the ship's maximum capacity of 3,600 people. In addition to handling procurement of equipment and materials and overseeing the construction, Carlisle designed the ship's lavish decor. Carlisle resigned Harland and Wolff in 1910 before the *Titanic's* completion. He was Lord Pirrie's brother-in-law, and he said that he found working with him difficult. So, he joined Welin Davit and Engineering Co. Ltd. in 1911. White Star had installed the davits but opted for one row of lifeboats instead of four. He was deeply unsettled by this.

Figure A.5. Alexander M. Carlisle.[6]

He gave evidence about the designed lifeboat capacity of *Titanic* at the British Board of Enquiry into the *Titanic* disaster in 1912.

Read more: http://channel.nationalgeographic.com/episode/rebuilding-*Titanic*-6451/Overview2#tab-*Titanic*-trio#ixzz1VVqfgYA4.

Captain E. J. Smith (1850–1912)

At the age of 13, Captain Smith went to Liverpool where he served his apprenticeship to begin a seafaring career. In 1880, he joined the White Star Line as the Fourth Officer of the *SS Celtic*. He served aboard the company's liners to Australia and to New York. In 1887, he served his first command on the *Republic*. In 1895, he progressed to the *Majestic* as captain for nine years, and he was called to transport troops to Cape Colony. He started to gain a reputation among passengers who would sail the Atlantic only in a ship under his command. In 1904, he commanded the line's newest ships on their maiden voyages and became known as the "Millionaires' Captain." He commanded the largest ship in the world, White Star's new *Baltic*, on her maiden voyage and, three years later, his second new "big ship," the *Adriatic*. His appointment to captain the Olympic-class ships was no surprise. Bruce Ismay believed he would help draw repeat passengers.

Figure A.6. Captain Smith.[7]

Portrayed by White Star as the most experienced captain and promoted to commander of the fleet, he was given the honor to captain *Titanic* on her maiden voyage. Responsible for the operation of the solution, his role began at the later stages of the project with the sea trials, as he and his officers took *Titanic* through her paces. They brought with them their nine months of experience with *Olympic* and six Atlantic crossings. However, in reality, he was haunted by a couple of incidents, notably *Olympic*'s collision with *HMS Hawke* that delayed *Titanic*'s maiden voyage by a month as she was repaired.

Captain Smith's actions contributed to *Titanic*'s disaster for several reasons. First, he was overconfident in the ship and its safety systems and viewed it as unsinkable. Second, his mistrust of technology swayed him from paying close enough attention to warning signs such as the Marconigram ice warnings. Third, during the maiden voyage, he failed to prevent a situation where Bruce Ismay took command of the ship, resulting in the ship racing toward Iceberg Alley at top speed. He was possibly the only person on board who could have countered Bruce Ismay. Captain Smith was lost with *Titanic*.

First Officer Murdoch (1873–1912)

Murdoch was born in Scotland in a notable line of Scottish seafarers who sailed the world's oceans. His father was Captain Samuel Murdoch, a master mariner and captain, as was his grandfather and four of his grandfather's brothers, very much a family tradition.

Figure A.7. First Officer Murdoch.[8]

Murdoch had worked through the White Star ranks to become one of the foremost senior officers. With 16 years of maritime experience, he was selected to be *Titanic*'s chief officer. With an "ordinary master's certificate," he built a reputation as a "canny and dependable man."

On the night of the collision, Murdoch was the officer in charge on the bridge. On receiving the warning, Murdoch gave an order of "Hard a'starboard" to avoid collision with ice, and then "Hard a'port." He almost succeeded in avoiding contact, but ended up grounding the ship. After the collision, he was put in charge of the starboard evacuation, during which he launched 10 lifeboats. Murdoch was last seen trying to free Collapsible A just before the bridge was submerged, and a huge wave washed him overboard.

Captain Lord of *Californian* (1877–1962)

Lord went to sea when he was 13 aboard the barque *Naiad*. In February 1901, at 23, Lord obtained his Master's Certificate, and three months later, obtained his Extra Master's Certificate. The company was taken over by the Leyland Line in 1900, but Lord continued service with the new company, and he was awarded his first command in 1906.

Lord was given full command of the SS *Californian* in 1911 and was sailing her on the night of April 14, 1912. He had avoided a major collision with ice by ordering full speed astern to stop on the edge of an extensive ice field. He then pro-actively instructed his radio operator Evans to send ice warnings to *Titanic*. Unfortunately, Evans was snubbed, and the *Californian*'s radio operator turned off his radio wireless and went to bed. At 12:45 a.m., the crew saw rockets from *Titanic,* but these were misunderstood. Captain Lord was informed at 1:10 a.m., but concluded the ship had stopped for the night, and all were having a party. No one on board the *Californian* tried to wake their wireless operator and ask him to contact the ship to ask why they fired rockets.

Figure A.8. Californian's captain, Stanley Lord.[9]

The conclusions of both the United States inquiry and the British inquiry seemed to disapprove of the actions of Captain Lord but stopped short of recommending charges. Lord resigned from the Leyland Line in August 1912, although some reports indicated he was dismissed. His inaction haunted him for the rest of his life as he tried to exonerate himself. His family has continued this to today.

Captain Rostron of *Carpathia* (1869–1940)

The *Carpathia* was on its regular route between New York City and Fiume when, on April 15, 1912, she received a distress signal from *Titanic*. Captain Rostron immediately ordered the ship to race toward the *Titanic*'s reported position, posting extra lookouts to help spot and maneuver around the ice he knew to be in the area. *Carpathia* was 58 nautical miles (93 km) from *Titanic's* position and the closest ship to respond to *Titanic*. *Carpathia* hit the maximum speed possible while traversing through dangerous ice floes. It took about 3½ hours to reach the *Titanic's* radioed position. Rostron prepared the ship for the survivors, getting blankets, food, and drinks ready, and putting his medical crew on standby to receive the injured survivors. Altogether, a list of 23 orders from Rostron to his crew was successfully implemented before *Carpathia* had even arrived at the scene of the disaster.

Figure A.9. Carpathia's captain, Arthur Henry Rostron.[10]

When Rostron believed he was getting close to the *Titanic*, he had green starburst rockets launched to encourage the

Titanic if she was still afloat, or her survivors, if she was not. *Carpathia* began to pick up survivors about an hour after the first starburst was seen by those in the lifeboats.

Captain Rostron won wide praise for his energetic efforts to reach the *Titanic* before she sank and his efficient preparations for and conduct of the rescue of the survivors. He was awarded a Congressional Gold Medal by the US Congress and, after World War I, was appointed Knight Commander of the Order of the British Empire. He was made the Commodore of the Cunard fleet before retiring in 1931.

Glossary and Abbreviations

Related to Project Management

BAC (Budget at Completion). The estimated total cost of the project when completed; the project baseline.

Change Control. The procedures used to identify, document, approve changes to the project baselines.

Change Management. The process for managing change in the project.

Chart of Accounts. A financial numbering system for monitoring project costs by category.

Code of Accounts. A numbering system for providing unique hierarchical identifiers for all items in the WBS.

Contingency and Management Reserves. Monies established and set aside for contingencies, which usually are not part of the cost performance baseline.

Contingency Plan. A response to a risk event only implemented if the risk event occurs.

Contingency Reserves or Buffers. Additional time set aside, regularly analyzed, and adjusted as data from the project becomes available.

Control Account. The management control point at which integration of scope, budget, and schedule takes place and at which performance is measured.

Cost Performance Baseline. The overall expected cost for the project when using a budget-at-completion (BAC) calculation. It is used to measure, monitor, and control overall cost performance on the project.

Crashing. A strategy to compress the project schedule without reducing project scope.

Critical path. The path with the longest duration within the project. It has the least float (usually no float). Task delayed on the critical path will delay the project.

Decomposition. A process that creates deliverables (in Create WBS process) or activities (in the Define Activities process).

Dependencies of activities:

- Discretionary—allow activities to happen in a preferred order because of best practices, conditions unique to the project work, or external events.

- External—outside control of the project.

- Mandatory—must be followed, and activities require a specific predefined order. These relationships are called hard logic.

EVM (Earned Value Management). A structured approach to planning, cost collection, and performance measurement. It integrates project scope, time, and cost objectives and the establishment of a baseline plan for performance measurement. There are three dimensions to EVM:

- **Planned Value (PV)** is the sum of the approved cost estimates for activities scheduled to be performed during a given period.

- **Actual Cost (AC)** is the amount spent in completing work in a given period.

- **Earned Value (EV)** is the sum of the approved cost estimates for activities completed during a given period.

These EVM dimensions produce important variables that are monitored during a project:

- **Cost Variance (CV)** is earned value (EV) minus actual cost (AC), i.e., $CV = EV - AC$; the difference between the budgeted cost of the work completed and the actual cost of completing the work; the **project is over budget if the number is negative.**

- **Schedule Variance (SV)** is earned value (EV) minus planned value (PV), i.e., $SV = EV - PV$; the difference between what was accomplished and what was scheduled; the **project is behind schedule if the number is negative.**

EAC (Estimate at Completion). The amount expected for the total project to cost on completion (at a particular point in the project).

Estimating:

- **Analogous**—using similar experience.

- **Parametric**—using formulas.

- **Three-Point**—for each activity, namely, optimistic, most likely, and pessimistic. An average is then created.

ETC (Estimate to Complete). The estimated additional costs to complete activities or the project.

Expert Judgment. Expertise, for example, with a particular type of hull design known as the "Belfast Bottom."

Fast Tracking. Project activities normally done sequentially are performed in parallel. It increases rework and project risk.

Float. The allowable time a scheduled activity can be delayed without delaying the end of the project.

Gold-plating. Over-elaboration of functions, incurring additional project costs.

Grade. The category or level of the characteristics of a product.

Lag time. The wait time before the next activity starts.

Lead time. The time subtracted from the downstream activity to bring successor activities closer to the start of the project. It is also the time saved by starting an activity before its predecessor finishes.

Leadership styles (from Verma, *Managing the Project Team*, pages 146–147) include the following:

- **Autocratic and Directing**, in which decisions are made solely by the project manager with little input from the team
- **Consultative Autocratic and Persuading**, in which decisions are still made only by the project manager with large amounts of input solicited from the team
- **Consensus and Participating**, in which the team makes decisions after open discussion and information gathering
- **Shareholder and Delegating** (otherwise known as laissez-faire or hands off), often considered a poor

leadership style in which the team has ultimate authority for final decisions, but little or no information exchange takes Place

Opportunities. Risk events or conditions favorable to the project.

Organizational Breakdown Structure (OBS). Organizational chart in which work package responsibility is related to the organizational unit responsible for performing that work.

Organizational Process Assets. Formal processes and best practices in shipbuilding.

Personal power (from Verma, *Human Resource Skills*, p. 233) includes the following:

- **Formal**—a legitimate form of power based on a person's position in the organization.

- **Reward**—a legitimate form of power based on positive consequences or outcomes the person can offer; it can also result from personal influence.

- **Coercive (Penalty)**—a legitimate form of power based on negative consequences or outcomes the person can inflict; it can also result from personal influence.

- **Referent**—a personal form of power based on a person's charisma or example as a role model (an earned power).

- **Expert**—a personal form of power based on the person's technical knowledge, skill, or expertise on some subject (an earned power).

Progressive Elaboration. The progressive improvement of a plan as more detailed information becomes available.

Project management. The process by which projects are defined, planned, monitored, controlled, and delivered such that the agreed benefits are realized. Projects are unique, transient endeavors undertaken to achieve a desired outcome. Projects bring about change, and project management is recognized as the most efficient way of managing such change.

Quality. The sum of the characteristics of the ship that meet the expectations of the project.

- **Quality Control**—involves measurement of the process or performance using quality control tools. It also compares and reports a project's actual progress with its standard.

- **Quality Assurance**—a process of regular structured reviews to ensure the project complies with the planned quality standards by using various collected measurements.

- **Quality Improvement**—the project performance is measured and evaluated, and corrective actions are applied to improve the product and the project.

- **Corrective Actions**—used to prevent unacceptable quality and improve the overall quality of the project management processes.

- **Prevention**—this is planned in a project and not inspected in, as it is always more cost effective to prevent mistakes than to correct them.

Resources. Equipment, people, subject matter experts, or materials.

Requirements:

- **Functional**—what a system does, namely, with ships, transportation and hospitality.

- **Nonfunctional**—all the requirements outside functional requirements. They define the operational characteristics of a system, or how a system delivers, and include performance, stability, security, maintainability, redundancy, and reliability.

- **Traceability Matrix**—records each requirement and tracks its attributes and changes throughout the project life cycle to highlight changes to the project scope.

Reserves - contingency and management. Contingency reserves are reserves of time or cost set aside to deal with the level of uncertainty that exists within the project. Management reserves, set by management, cover unknown risks.

Risk Management:

- **Risk Identification**—methodical, planned approach to identifying risks.

- **Risk Analysis**—qualitative is where risks are scored and ranked based on their probability and impact. Quantitative assesses numerically the probability and impact of the identified risks and creates a risk score for the project.

- **Risk response planning**—decreases the possibility of risks affecting the project adversely and maximizes positive risks to help the project by assigning responsibilities to project team members close to the risk event.

Schedule compression. The project schedule is shortened without changes to the project scope through techniques such as Crashing and Fast Tracking.

Scope Baseline. The approved project scope statement along with the WBS (and WBS dictionary).

SLA (Service-Level Agreements). Agreements between parties (supplier and client) for levels of service.

SLO (Service-Level Objectives). Objectives set for levels of service to reach.

Sponsor. Typically, an initiator of the project who is brokering the funding for it.

Stakeholder for the project. Anyone associated in having a stake in the project, identified using a stakeholder map, and held in a stakeholder register, a directory of all stakeholders. The expectations of the stakeholders should be identified as well as their stance on the project (supporter, negative, neutral).

Sunk Costs. Investments already made and hence not included when determining alternatives.

Threats. Risk events unfavorable to the project.

Total Float. Determines flexibility in the schedule. It is measured by subtracting the early dates from late dates, along the critical path of completion.

VAC (Variance at Completion). The difference between the total amount the project was supposed to cost (BAC) and the amount the project is now expected to cost (EAC).

WBS (Work Breakdown Structure). A framework for defining the total scope of project work into smaller, more manageable pieces.

- **WBS dictionary**—defines the WBS components.
- **Work Package**—the lowest level of a WBS; where cost estimates are made.

Examples of Functional Requirements

Functional Requirements: what a system does, namely, with ships, transportation and hospitality.

Nonfunctional Requirements: all the requirements outside functional requirements.

- Functional defines what the system needs to do:
  - Business Rules
  - Transaction corrections, adjustments, cancellations
  - Administrative functions
  - Authentication
  - Authorization-functions user is delegated to perform
  - Audit tracking
  - External interfaces
  - Certification requirements
  - Reporting requirements
  - Historical data
  - Legal or regulatory requirements
- Non-functional defines how the system needs to behave:
  - Performance-response time, throughput, utilization, static volumetric
  - Scalability
  - Capacity
  - Availability
  - Reliability
  - Recoverability
  - Maintainability

- ○ Serviceability
- ○ Security
- ○ Regulatory
- ○ Manageability
- ○ Environmental
- ○ Data integrity
- ○ Usability
- ○ Interoperability

Functional Requirements and Nonfunctional Requirements related to a ship:

- Functional-what the system needs to do:
  - ○ Deliver customer experience
    - Time of crossing (Wednesday ship)
    - Hospitality-space per passenger (cabins, functional areas)
    - Dining (Restaurants, cafes, bars)
    - Entertainment (Galas, balls, library, smoking rooms, games)
    - Exercise and fitness (Gymnasium, squash court, swimming pool, Turkish baths)
    - Services (Post office, mail, telegrams, barbers)
  - ○ Meet certification and regulatory requirements
- Non-functional-how the system needs to behave:
  - ○ To meet SLA's and government regulations
  - ○ Dimensions of the ship (height, length, width)
    - Displacement
    - Weight and its distribution

- Handling characteristics
  - Speed (performance and spare capacity) number of engines
  - Power (HP) of engines, scalability, redundancy and reliability
  - Number of furnaces, boilers, and coal bunkers
  - Number of propellers, their size
  - Steering and rudder size
- Safety requirements
- Regulatory

Bibliography

[1] Allen, F. L. *The Great Pierpont Morgan*. New York: Marboro Books, 1949.

[2] Bonsall, Thomas E. *Great Shipwrecks of the 20th Century*. New York: Gallery Books, 1988.

[3] Bristow, Diana. Titanic: *Sinking the Myths*. KatCo Literary Group of Central California, June 1995.

[4] Brown, David. *The Last Log of the Titanic*. McGraw-Hill, 2002.

[5] Bullock, Shan. "Thomas Andrews: His Apprenticeship at Harland and Wolff, in *Thomas Andrews Shipbuilder 1912*. University of Toronto Libraries, 2011. <http://www.libraryireland.com/Thomas-Andrews-Shipbuilder/Apprenticeship-Harland-Wolff.php>.

[6] Davie, Michael. *The* Titanic: *The Full Story of a Tragedy*. The Bodleyhead Ltd., 1986.

[7] Eaton, John P., and Charles A. Haas. Titanic: *Triumph and Tragedy*. Norton, 1995.

[8] Hoyt, E. P. *The House of Morgan*. Mead, NY: Dodd, 1966.

[9] Hyslop, Donald, Alastair Forsyth, and Sheila Meminia. Titanic *Voices*. New York: St. Martin's Press, 1998.

[10] Lord, Walter. *A Night to Remember*. New York: Holt, Rinehart & Winston, 1955.

[11] Lord, Walter. *The Night Lives On*. New York: Holt, Rinehart & Winston, 1985.

[12] Maxtone-Graham, John. *The Only Way to Cross*. Barnes & Noble Books, 1998.

[13] *Titanic* - The Ship Magnificent. http://www.*Titanic*-theshipmagnificent.com

[14] *Titanic* Inquiry Project—Electronic copies of British and American. Fully Searchable transcripts of the complete US Senate and British Board of Trade inquiries and reports into the sinking of the *Titanic*. http://www.titanicinquiry.org/

[15] Wade, Wyn Craig. *The* Titanic*: End of a Dream*. New York: Rawson-Wade, 1979.

[16] Wels, Susan. *Titanic*: Legacy of the World's Greatest Ocean Liner. Alexandria, VA: Time-Life Books, 2000.

Endnotes

Chapter One

[1] This image (or other media file) is in the public domain because its copyright has expired.

[2] *New York Times*, June 9, 1924.

[3] Lord Pirrie and Bruce Ismay inspecting ship. This photograph was used courtesy of the Ulster Folk & Transport Museum.

[4] This image (or other media file) is in the public domain because its copyright has expired. Source: Fashion plate of the upper class in front of Harrods, 1909.

[5] Illustration of the IMM Share Issue. This image (or other media file) is in the public domain because its copyright has expired.

[6] 1902 editorial cartoon in *Puck*. This image (or other media file) is in the public domain because its copyright has expired.

Chapter Two

[1] Jennifer Hooper McCarty and Tim Foecke, *What Really Sank the Titanic: New Forensic Discoveries* (New York, NY: Citadel Press, 2008).

[2] C. Knick Harley, "Aspects of the Economics of Shipping," in L. R. Fischer and G. E. Panting, "Change and Adaptation in Maritime History: The North Atlantic Fleets in the Nineteenth Century," Maritime History Group, Memorial University of Newfoundland, *Proceedings of the Sixth Conference of the Atlantic Canada Shipping Project,* April 1–3, 1982, p. 176.

[3] This image (or other media file) is in the public domain because its copyright has expired. Photograph of Immigrants arriving at Ellis Island.

[4] Courtesy of the National Archives of the United Kingdom: ref. BT 100/259.

[5] Courtesy of the National Archives of the United Kingdom: ref. BT 100/259.

[6] The evolution in ship size. Courtesy of the *New York Times* archives.

[7] This image (or other media file) is in the public domain because its copyright has expired.

[8] H1227 Construction of the massive slipway. This photograph was used courtesy of the Ulster Folk & Transport Museum.

[9] This image (or other media file) is in the public domain because its copyright has expired.

Chapter Three

[1] The drawing office at the Belfast shipyards. This photograph was used courtesy of the Ulster Folk & Transport Museum.

[3] H2401 April-1910 builder's case model of *Olympic/Titanic* with sign—the largest ship in the world. This photograph was used courtesy of the Ulster Folk & Transport Museum.

[4] This image (or other media file) is in the public domain because its copyright has expired.

[5] This image (or other media file) is in the public domain because its copyright has expired. Title US inquiry into the loss of the *Titanic*: longitudinal section and plans showing bulkheads.

[6] H1471 Electric vertical-sliding watertight door of *Olympic*. This photograph was used courtesy of the Ulster Folk & Transport Museum.

[7] This image (or other media file) is in the public domain because its copyright has expired.

[8] How wireless worked. Newspaper unknown. This image (or other media file) is in the public domain because its copyright has expired.

[9] Photo #NH 63063 *S.S. Arizona* (British Passenger Steamer, 1879). This photograph was used courtesy of the US Naval Historical Center Photograph.

[10] This image (or other media file) is in the public domain because its copyright has expired. Source: *Wreck and Sinking of the Titanic* edited by Marshall Everett, a pseudonym of W. H. Walter, the name given for the copyright. Published in the USA in 1912.

[11] This image (or other media file) is in the public domain because its copyright has expired.

[12] This image (or other media file) is in the public domain because its copyright has expired.

[13] Source: British Board of Trade report on the disaster.

[14] This image (or other media file) is in the public domain because its copyright has expired. Title: 1st class, grand dining saloon, seates [sic] 500--OLYMPIC STEAMSHIP. Date Created/Published: [between 1911 and 1920]. Reproduction Number: LC-USZ62-99340 (b&w film copy neg.). Rights Advisory: No known restrictions on publication. Call Number: LOT 11261 <item> [P&P].

[15] Engineering May 11. This image (or other media file) is in the public domain because its copyright has expired.

Chapter Four

[1] Mold loft. This image (or other media file) is in the public domain because its copyright has expired.

[2] This image (or other media file) is in the public domain because its copyright has expired. Title: Apprentice School, First day of class, 1915, US Navy Yard, Mae Island, CA, ca. 1915. ARC Identifier 296836 Item from Record Group 181: Records of Naval Districts and Shore Establishments, 1784–2000. Rights Advisory: No known restrictions on reproduction. The US National Archives and Records Administration.

[3] Olympics keel is laid. This photograph was used courtesy of the Ulster Folk & Transport Museum.

[4] Advertising brochure. This image (or other media file) is in the public domain because its copyright has expired.

[5] Source: Jennifer Hooper McCarty and Tim Foecke, *What Really Sank the Titanic*, 2009.

[6] Constructing the ten decks. This photograph was used courtesy of the Ulster Folk & Transport Museum.

[7] Source: Jennifer Hooper McCarty and Tim Foecke, *What Really Sank the Titanic*, 2009.

[8] Stockyard holding the thousands of steel plates. This image (or other media file) is in the public domain because its copyright has expired. Title: OLYMPIC and TITANIC—view of bows [in shipyard construction scaffolding]. Date Created/Published: [between 1909 and 1911]. Reproduction Number: LC-USZ62-67359 (b&w film copy neg.) Rights Advisory: No known restrictions on publication. Call Number: LOT 11261 [item] [P&P]. Repository: Library of Congress Prints and Photographs Division Washington, D.C. 20540 USA.

[9] Courtesy of the National Archives of the United Kingdom: ref. BT 100/259.

[10] Courtesy of the National Archives of the United Kingdom: ref. BT 100/259.

[11] Courtesy of the National Archives of the United Kingdom: ref. BT 100/259.

[12] Courtesy of the National Archives of the United Kingdom: ref. BT 100/259.

[13] Boilers laid out. This image (or other media file) is in the public domain because its copyright has expired.

[14] Source: Jennifer Hooper McCarty and Tim Foecke, *What Really Sank the Titanic,* 2009.

[15] Vision of the promenade deck. This image (or other media file) is in the public domain because its copyright has expired.

[16] 1910 editorial cartoon in *Puck*. This image (or other media file) is in the public domain because its copyright has expired.

[17] *Olympic* & *Titanic* in slipways Queen's Yard. WAG 2079. This photograph was used courtesy of the Ulster Folk & Transport Museum.

[18] This image (or other media file) is in the public domain because its copyright has expired. RMS Olympic launching. Date 20 October 1910(1910-10-20). Source http://www.bytenet.net/rmscaronia/Main%20 Olympic%20Page.htm, Author Robert Welsh (died 1936).

[19] This image (or other media file) is in the public domain because its copyright has expired. Courtesy of *Scientific American* magazine.

[20] Lifting a boiler on board. 400-79. This photograph was used courtesy of the Ulster Folk & Transport Museum.

[21] H1996 Fitting engines in the engine room. This photograph was used courtesy of the Ulster Folk & Transport Museum.

[22] This image (or other media file) is in the public domain because its copyright has expired.

[23] 1st-Class Dining Saloon on D-Deck of the *Olympic*. This image (or other media file) is in the public domain because its copyright has expired.

[24] Wheel House Telephones. This image (or other media file) is in the public domain because its copyright has expired.

[25] Engine room telephones. This image (or other media file) is in the public domain because its copyright has expired.

[26] H1557. This photograph was used courtesy of the Ulster Folk & Transport Museum.

[27] H1508 Onboard sternward view of upper deck and lifeboats. This photograph was used courtesy of the Ulster Folk & Transport Museum.

[28] This media file is in the public domain in the United States. This applies to US works where the copyright has expired, often because its first publication occurred prior to January 1, 1923.

[29] This image (or other media file) is in the public domain because its copyright has expired. Date Created/ Published: 1912 April 15. Reproduction Number: LC-USZ62-26743 (b&w film copy neg.). Rights Advisory: No known restrictions on reproduction. Call Number: LOT 6668 [item] [P&P]. Repository: Library of Congress Prints and Photographs Division Washington, D.C. 20540 USA.

[30] H1568 Starboard stern view of completed ship. This photograph was used courtesy of the Ulster Folk & Transport Museum.

[31] Admission pass. This image (or other media file) is in the public domain because its copyright has expired.

[32] This image (or other media file) is in the public domain because its copyright has expired.

[33] Posters. This image (or other media file) is in the public domain because its copyright has expired.

[34] IMM advertisement. This image (or other media file) is in the public domain because its copyright has expired.

Chapter Five

[1] Source: Donald Hyslop, Alastair Forsyth, Sheila Jemima, *Titanic Voices*, 1997.

[2] This image (or other media file) is in the public domain because its copyright has expired. Title: OLYMPIC—maiden voyage. Date Created/Published: [between 1909 and 1911]. Summary: Side view, with bow to right, launched at Belfast, Northern Ireland. Reproduction Number: LC-USZ62-76281 (b&w film copy neg.). Rights Advisory: No known restrictions on publication. Call Number: LOT 11261 [item] [P&P]. Repository: Library of Congress Prints and Photographs Division Washington, D.C. 20540 USA.

[3] This image (or other media file) is in the public domain because its copyright has expired. Title: Photograph of a ship deck that was similar to the Titanic, 04/12/1912, ARC Identifier 278332/MLR Number 383. Item from Recond Group 21: Records of District Courts of the United States 1685-2004. Rights Advisory: No known restrictions on publication. The US National Archives and Records Administration.

[4] This image (or other media file) is in the public domain because its copyright has expired. Title: OLYMPIC. Creator(s): Bain News Service, publisher. Date Created/Published: [1911 June 6]. Reproduction Number: LC-DIG-ggbain-09362 (digital file from original negative). Rights Advisory: No known restrictions on publication. Call Number: LC-B2- 2218-2 [P&P] LOT 11261 (Corresponding print). Repository: Library of Congress Prints and Photographs Division Washington, D.C. 20540 USA http://hdl.loc.gov/loc.pnp/pp.print.

[5] This image (or other media file) is in the public domain because its copyright has expired. Courtesy of *Scientific American* magazine.

[6] Period picture postcard. This image (or other media file) is in the public domain because its copyright has expired.

[7] Titanic fitting out. This photograph was used courtesy of the Library of Congress Prints and Photographs Division, Washington, D.C. 20540 USA. Reproduction Number: LC-USZ62-56585 (b&w film copy neg.) Rights Advisory: No known restrictions on reproduction. Call Number: LOT 6668 [item] [P&P]

[8] Period picture postcard. This image (or other media file) is in the public domain because its copyright has expired.

[9] H1622. This photograph was used courtesy of the Ulster Folk & Transport Museum.

[10] Period picture postcard. This image (or other media file) is in the public domain because its copyright has expired.

[11] Courtesy of the National Archives of the United Kingdom: ref. ADM 116/1163.

[12] Courtesy of the National Archives of the United Kingdom: ref. ADM 116/1163.

[13] Courtesy of the National Archives of the United Kingdom: ref. ADM 116/1163.

[14] Source: Admiralty report, National Archives of the United Kingdom: ref. ADM 116/1163.

[15] Source: Jennifer Hooper McCarty and Tim Foecke, *What Really Sank the Titanic*, 2009.

[16] "Liner Olympic Hits Wreck", *Chicago Daily News*, by The Associated Press, Tuesday, February 27, 1912, p. 3.

[17] This image (or other media file) is in the public domain because its copyright has expired. Title: [View of the OLYMPIC in dry dock]. Date Created/Published: [ca. 1911]. Reproduction Number: LC-USZ62-34781 (b&w film copy neg.). Rights Advisory: No known restrictions on reproduction. Call Number: LOT 6668 [item] [P&P]. Repository: Library of Congress Prints and Photographs Division Washington, D.C. 20540 USA.

[18] The Launch of a Levithan," The Irish Times, Saturday, May 28, 1911.

[19] Courtesy of the National Archives of the United Kingdom: ref. BT 100/259.

[20] This image (or other media file) is in the public domain because its copyright has expired. *Titanic* beginning a day of sea trials, April 2, 1912. (NARA, RG 306, Records of the US Information Agency). Rights Advisory: No known restrictions on reproduction.

Chapter Six

[1] This image (or other media file) is in the public domain because its copyright has expired.

[2] Period picture postcard. This image (or other media file) is in the public domain because its copyright has expired.

[3] Source: David G. Brown, *The Last Log of the Titanic*, 2001.

[4] *New York Times*. This image (or other media file) is in the public domain because its copyright has expired.

[5] Period advertisement. This image (or other media file) is in the public domain because its copyright has expired.

[6] This media file is in the public domain in the United States. This applies to US works where the copyright has expired, often because its first publication occurred prior to January 1, 1923.

[7] *Daily Mirror*. This image (or other media file) is in the public domain because its copyright has expired.

[8] This image (or other media file) is in the public domain because its copyright has expired.

Chapter Seven

[1] Reuters. This image (or other media file) is in the public domain because its copyright has expired.

[2] Source: http://www.telegraph.co.uk/culture/books/8677437/
Titanic-builder-J-Bruce-Ismay-doomed-the-moment-he-
jumped-ship.html.

[3] This image (or other media file) is in the public domain
because its copyright has expired. MIKAN 2843910:
The Ill-Fated S.S. Titanic. Box PC-17 Item no. assigned
by LAC 260 90: Open Archival reference no. R13349-
260-0-E. Government of Canada and Library and
Archives Canada.

[4] This image (or other media file) is in the public domain
because its copyright has expired.

[5] This image (or other media file) is in the public domain
because its copyright has expired. Summary: The great
Titanic disaster, with wireless operator on shipboard
receiving distress call. Reproduction Number:: LC-
USZ62-90833 (b&w film copy neg.). Rights Advisory:
No known restrictions on publication. Repository:
Library of Congress Prints and Photographs Division,
Washington, D.C. 20540 USA.

[6] Shipbuilders Model of *Titanic*, Courtesy of Liverpool
Maritime Museum.

[7] Source: Read more: http://www.dailymail.co.uk/news/
article-478269/Is-man-sank-Titanic-walking-vital-
locker-key.html#ixzz1WeXDoNsx.

[8] This image (or other media file) is in the public domain
because its copyright has expired.

Chapter Eight

[1] This image (or other media file) is in the public domain
because its copyright has expired.

[2] Forecast telephone. This image (or other media file) is in the
public domain because its copyright has expired.

[3] Shipbuilders Model of *Titanic*, Courtesy of Liverpool
Maritime Museum.

[4] Watertight door indicator panel. This image (or other media file) is in the public domain because its copyright has expired.

Chapter Eleven

[1] This image (or other media file) is in the public domain because its copyright has expired.

[2] This image (or other media file) is in the public domain because its copyright has expired.

Chapter Twelve

[1] This image (or other media file) is in the public domain because its copyright has expired. US Navy daily memorandum reporting the Titanic's collision with an iceberg, April 15, 1912. (National Archives-Northeast Region, New York City, RG 21, Records of District Courts of the United States).

[2] This image (or other media file) is in the public domain because its copyright has expired.

[3] This image (or other media file) is in the public domain because its copyright has expired. Title: TITANIC lifeboats on way to CARPATHIA. Creator(s): Bain News Service, publisher. Date Created/Published: [1912 April]. Summary: Photo related to the disaster of the RMS TITANIC, which struck an iceberg in April 1912 and sank, killing more than 1,500 people. Photo possibly taken by passenger J. W. Barker. (Source: Flickr Commons project, 2008). Reproduction Number: LC-DIG-ggbain-11212 (digital file from original negative). Rights Advisory: No known restrictions on publication. Call Number: LC-B2-2485-4 [P&P]. Repository: Library of Congress Prints and Photographs Division, Washington, D.C. 20540 USA.

[4] Title: last lifeboat successfully launched from Titanic. Date Created/Published: [15 April 1912]. Summary: Photo possibly taken from Carpathia, the ship that

received the Titanic's distress signal and came to rescue the survivors. Source: http://www.archives. gov/publications/the_record/march_1998/titanic.html Reproduction Number.

[5] This image (or other media file) is in the public domain because its copyright has expired. Title: Lifeboat. Photograph above was taken by a passenger of the Carpathia, the ship that received the Titanic's distress signal and came to rescue the survivors. 04/12/1912. ARC Identifier 278331 / MLR Number 383 (National Archives–Northeast Region, New York City, RG 21, Records of District Courts of the United States). No known restrictions on reproduction. The US National Archives and Records Administration.

[6] This image (or other media file) is in the public domain because its copyright has expired. Title: At White Star office after TITANIC disaster. Creator(s): Bain News Service, publisher. Date Created/Published: [1912 April] (date created or published later by Bain). Reproduction Number: LC-DIG-ggbain-10352 (digital file from original neg.). Rights Advisory: No known restrictions on publication. Call Number: LC-B2- 2391- 29 [P&P]. Repository: Library of Congress Prints and Photographs Division, Washington, D.C. 20540 USA.

Chapter Thirteen

[1] This image (or other media file) is in the public domain because its copyright has expired.

[2] This image (or other media file) is in the public domain because its copyright has expired. Title: TITANIC lifeboat being hoisted to drain it of water: 04/12/1912. ARC Identifier 278331 / MLR Number 383. ARA's Northeast Region (New York City), New York, NY. Rights Advisory: No known restrictions on reproduction. The US National Archives and Records Administration.

³ *The New York Times* archives. This image (or other media file) is in the public domain because its copyright has expired.

⁴ *The Denver Times.* This image (or other media file) is in the public domain because its copyright has expired.

⁵ *The London Herald.* This image (or other media file) is in the public domain because its copyright has expired.

⁶ *The Washington Post.* This image (or other media file) is in the public domain because its copyright has expired.

⁷ This image (or other media file) is in the public domain because its copyright has expired.

⁸ This image (or other media file) is in the public domain because its copyright has expired.

⁹ The Evening Gazette. This image (or other media file) is in the public domain because its copyright has expired.

¹⁰ The Times Dispatch. This image (or other media file) is in the public domain because its copyright has expired.

¹¹ This image (or other media file) is in the public domain because its copyright has expired. Title: [Groups of TITANIC survivors aboard rescue ship CARPATHIA: unidentified group on deck]. Date Created/ Published: c1912 May 27. Reproduction Number: LC-USZ62-56453 (b&w film copy neg.). Rights Advisory: No known restrictions on publication. Call Number: LOT 10630 [item] [P&P]. Repository: Library of Congress Prints and Photographs Division, Washington, D.C. 20540 USA.

¹² This image (or other media file) is in the public domain because its copyright has expired. Reproduction Number: LC-USZ62-26635 (b&w film copy neg.) Rights Advisory: No known restrictions on reproduction. Call Number: LOT 6668 [item] [P&P] Repository: Library of Congress Prints and Photographs Division, Washington, D.C. 20540 USA.

Call Number: BIOG FILE - Brown, Mrs. J. J. <item>
[P&P] [P&P]. Repository: Library of Congress Prints
and Photographs Division, Washington, D.C. 20540
USA.

[17] This image (or other media file) is in the public domain
because its copyright has expired.

[18] This image (or other media file) is in the public domain
because its copyright has expired.

[19] This image (or other media file) is in the public domain
because its copyright has expired.

[20] This image (or other media file) is in the public domain
because its copyright has expired. Title: [TITANIC
disaster. Senate Investigating Committee questioning
individuals at the Waldorf Astoria]. Date Created/
Published: c1912 May 27. Summary: J. Bruce Ismay
at end of table, being questioned. Reproduction
Number: LC-USZ62-68081 (b&w film copy neg.).
Rights Advisory: No known restrictions on publication.
Call Number: LOT 10630 [item] [P&P]. Repository:
Library of Congress Prints and Photographs Division,
Washington, D.C. 20540 USA.

[21] This image (or other media file) is in the public domain
because its copyright has expired. Title: [TITANIC
disaster. Senate Investigating Committee questioning
individuals at the Waldorf Astoria]. Date Created/
Published: c1912 May 27. Summary: J. Bruce
Ismay being questioned. Reproduction Number:
LCUSZ62-68079 (b&w film copy neg.) Rights
Advisory: No known restrictions on publication.
Call Number: LOT 10630 [item] [P&P]. Repository:
Library of Congress Prints and Photographs Division,
Washington, D.C. 20540 USA.

Chapter Fourteen

[1] This image (or other media file) is in the public domain
because its copyright has expired. Title: Fleet,
Frederick. Titanic. Creator(s): Harris & Ewing,

photographer. Date Created/Published: 1912. Reproduction Number: LC-DIG-hec-00939 (digital file from original negative). Rights Advisory: No known restrictions on publication. Call Number: LC-H261-1168 [P&P]. Repository: Library of Congress Prints and Photographs Division, Washington, D.C. 20540 USA.

[2] This image (or other media file) is in the public domain because its copyright has expired. Title: [TITANIC disaster. Senate Investigating Committee questioning individuals at the Waldorf Astoria]. Date Created/Published: c1912 May 29. Summary: Harold Bride, wireless operator. Reproduction Number: LC-USZ62-68078 (b&w film copy neg.). Rights Advisory: No known restrictions on publication. Call Number: LOT 10630 [item] [P&P]. Repository: Library of Congress Prints and Photographs Division, Washington, D.C. 20540 USA.

[3] This image (or other media file) is in the public domain because its copyright has expired. Title: [TITANIC disaster. Senate Investigating Committee questioning individuals at the Waldorf Astoria]. Date Created/Published: c1912 May 29. Summary: Harold Thomas Coffam, wireless operator. Reproduction Number: LC-USZ62-68080 (b&w film copy neg.). Rights Advisory: No known restrictions on publication. Call Number: LOT 10630 [item] [P&P]. Repository: Library of Congress Prints and Photographs Division, Washington, D.C. 20540 USA.

[4] This image (or other media file) is in the public domain because its copyright has expired.

[5] This image (or other media file) is in the public domain because its copyright has expired.

[6] This image (or other media file) is in the public domain because its copyright has expired.

[7] William B. Saphire, The White Star Line and the International Mercantile Marine Company.

Chapter Fifteen

[1] According to "Chaos, a recipe for success," Standish Group, 1994–2010, only 26 percent of all projects today finish on time, on budget, and with all the features and functions originally specified.

Appendix A

[1] This image (or other media file) is in the public domain because its copyright has expired.

[2] This image (or other media file) is in the public domain because its copyright has expired.

[3] This image (or other media file) is in the public domain because its copyright has expired. J. P. Morgan photo from Images of American Political History, public domain. This media file is in the public domain in the United States. This applies to US works where the copyright has expired, often because its first publication occurred prior to January 1, 1923.

[4] J. Bradford De Long, "J. P. Morgan and His Money Trust," Harvard University and National Bureau of Economic Research1, September 1991.

[5] This image (or other media file) is in the public domain because its copyright has expired.

[6] This image (or other media file) is in the public domain because its copyright has expired.

[7] This image (or other media file) is in the public domain because its copyright has expired.

[8] This image (or other media file) is in the public domain because its copyright has expired.

[9] This image (or other media file) is in the public domain because its copyright has expired.

[10] Title: Capt. Rostron of the Carpathia. Creator(s): Bain News Service, publisher. Date Created/Published: [1912 April]. Summary: Arthur Henry Rostron, captain of

the Carpathia, full-length portrait, standing, facing front. Reproduction Number: LC-USZ62-121011 (b&w film copy neg.). Rights Advisory: No known restrictions on publication. Call Number: LC-B2- 2485-4 [P&P]. Repository: Library of Congress Prints and Photographs Division Washington, D.C. 20540 USA.

Index

L

M

N

O

P

Q

R

S

T

Taft, President 271-272, 280

Testing *(See Project phases, testing)*

Third class 19, 20, 22, 28, 30-31, 40, 71, 194, 228, 243, 254, 297, 304-308

Time Management

Estimate duration 34-35

Critical path 103-104, 157, 160-161, 165, 340, 346

Milestones 137

Schedule 23, 27, 32, 34, 36-38, 46, 55-56, 59, 61, 66-67, 92, 100, 103-104, 123, 142, 157, 159-165, 173-178, 340-342, 345-346; crashing (with outside labor) 160, 165, 340, 345; fast tracking 161, 342, 345;

Sequence of activities 34

Time dependencies 103, 340

U

User Outage Minute *(See Outage)*

W

Water pumps 75, 230-231, 304

Watertight (compartments, doors, tank) 41, 75-79, 86, 91, 102-103, 107, 135, 150, 155, 222-223, 231, 237, 241, 253, 292, 313

Wilding, Edward (Andrews' deputy) 71

Workforce (project) *(See also Human Resource Management)*

About the Author

 As the founder behind the "Lessons from History" series, Mark Kozak-Holland brings years of experience as a consultant who helps Fortune 500 companies formulate projects that leverage emerging technologies. Since 1985, he has straddled the business and Information Technology (IT) worlds, making these projects happen. He is a Project Management Professional (PMP), IPMA-D, a certified business consultant, the author of several books, and a noted speaker. As a historian, Kozak-Holland seeks the wisdom of the past to help others avoid repeating mistakes and to capture time-proven techniques. His lectures have been very popular at gatherings of project managers and CIOs.

Mark is very passionate about history and sees its potential use as an educational tool in business today. As a result, he has been developing the "Lessons from History" series for organizations, applying today's IT to common business problems. It is written for primarily business and IT professionals looking for inspiration for their projects. It uses relevant historical case studies to examine how historical projects and emerging technologies of the past solved complex problems.

For thousands of years, people have run projects that leveraged emerging technologies of the time to create unique and wonderful structures such as the pyramids, buildings, or bridges. Similarly, people have gone on great expeditions and journeys and raced their rivals in striving to be first, for example, circumnavigating the world or conquering the Poles. These were all forms of projects that required initiation, planning and design, production, implementation, and breakout.

The series looks at historical projects and then draws comparisons to challenges encountered in today's projects. It outlines the stages involved in delivering a complex project, providing a step-by-step guide to the project deliverables. It vividly describes the crucial lessons from historical projects and complements these with some of today's best practices, which makes the whole learning experience more memorable.

The series should inspire the reader, as these historical projects were achieved with a less sophisticated emerging technology.

E-mail: mark.kozak-holl@sympatico.ca

Web Sites: http://www.mmpubs.com/kozak-holland/
http://www.lessons-from-history.com/

Did you like this book?

If you enjoyed this book, you will find more interesting books at

www.MMPubs.com

Please take the time to let us know how you liked this book. Even short reviews of 2-3 sentences can be helpful and may be used in our marketing materials. If you take the time to post a review for this book on Amazon.com, let us know when the review is posted and you will receive a free audiobook or ebook from our catalog. Simply email the link to the review once it is live on Amazon.com, with your name, and your mailing address—send the email to orders@mmpubs.com with the subject line "Book Review Posted on Amazon."

If you have questions about this book, our customer loyalty program, or our review rewards program, please contact us at info@mmpubs.com.

Multi-Media Publications Inc.

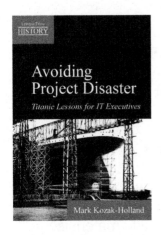

Avoiding Project Disaster: Titanic Lessons for IT Executives

Imagine you are in one of *Titanic's* lifeboats. As you look back at the wreckage, you wonder what could have happened. What were the causes? How could things have gone so badly wrong?

Titanic's maiden voyage was a disaster waiting to happen as a result of the compromises made in the project that constructed the ship. This book explores how modern executives can take lessons from a nuts-and-bolts construction project like *Titanic* and use those lessons to ensure the right approach to developing online business solutions.

Avoiding Project Disaster is about delivering IT projects in a world where being on time and on budget is not enough. You also need to be up and running around the clock for your customers and partners. This book will help you successfully maneuver through the ice floes of IT management in an industry with a notoriously high project failure rate.

ISBN: 9781895186734   (paperback)

http://www.mmpubs.com/disaster

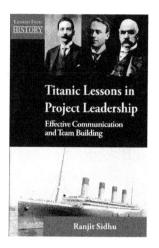

Titanic Lessons in Project Leadership: Effective Communication and Team Building

In this book, we see how "small" and easily overlooked behavioral and communication issues can aggregate through a project to become seemingly unthinkable errors.

It is critical that project managers and leaders have the skills to deal effectively with people issues. You need to be just as comfortable managing conflict and motivating your team as you are with planning your work and conducting a risk analysis. When faced with challenging deadlines and the pressures that go with managing projects, it is easy just to focus on getting the task done. This is most likely at the expense of having those difficult conversations with upset stakeholders and disgruntled customers; the people who ultimately determine whether the project is a success or failure.

This book focuses on the people aspects of the Titanic story; the key stakeholders, power dynamics, underlying perceptions, communication, leadership and team interactions. Ranjit Sidhu draws on this tragic tale to focus on the "behind the scenes" aspects of human communication and leadership to guide you in the right direction for making that vital difference to your current projects.

Combining contemporary management theory with her own insights and extensive project management experience, Ranjit offers practical guidance and lessons from history that will help you gain a deeper understanding of how leaders and teams can operate at their very best.

ISBN: 9781554891207   (paperback)

http://www.mmpubs.com/

Project Lessons from The Great Escape (Stalag Luft III)

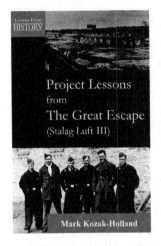

While you might think your project plan is perfect, would you bet your life on it?

In World War II, a group of 220 captured airmen did just that – they staked the lives of everyone in the camp on the success of a project to secretly build a series of tunnels out of a prison camp their captors thought was escape proof.

The prisoners formally structured their work as a project, using the project organization techniques of the day. This book analyzes their efforts using modern project management methods and the nine knowledge areas of the *Guide to the Project Management Body of Knowledge* (PMBoK).

Learn from the successes and mistakes of a project where people really put their lives on the line.

ISBN: 9781895186802   (paperback)

http://www.mmpubs.com/escape

Polaris: Lessons in Risk Management

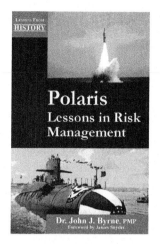

Risk management is one of the most important practices that a manager can employ to help drive a successful outcome from a project. Good risk management allows organizations to proactively respond to risks.

Unfortunately, many managers believe risk management to be too time consuming or too complicated. Some find it to be shrouded in mystery.

This book is designed to demystify risk management, explaining introductory and advanced risk management approaches in simple language. To illustrate the risk management concepts and techniques, this book uses real-life examples from a very influential project that helped change the course of world history -- the project that designed and built the Polaris missile and accompanying submarine launch system that became a key deterrent to a Soviet nuclear attack during the Cold War. The Polaris design and construction project employed many risk management approaches, inventing one that is still widely used today.

Containing a foreword by James R. Snyder, one of the founders of the Project Management Institute (PMI), this book is structured to align with the risk management approach described in PMI's the Project Management Body of Knowledge (PMBOK Guide).

ISBN: 9781554890972   (paperback)

The History of Project Management

The Pyramid of Giza, the Colosseum, and the Transcontinental Railroad are all great historical projects from the past four millennia. When we look back, we tend to look at these as great architectural or engineering works. Project management tends to be overlooked, and yet its core principles were used extensively in these projects.

This book takes a hard look at the history of project management and how it evolved over the past 4,500 years. It shows that "modern" project management practices did not just appear in the past 100 years but have been used — often with a lot of sophistication — for thousands of years.

Examining archaeological evidence, artwork, and surviving manuscripts, this book provides evidence of how each of the nine knowledge areas of project management (as shown in PMI's PMBoK® Guide) have been practiced throughout the ages. The book covers the period from the construction of the ancient pyramids up to the 1940s. A future companion volume will cover more recent developments during the war years and the technology boom leading up to the present.

As readers explore the many case studies in this book, they will discover fascinating details of innovative projects that produced many of our most famous landmarks and voyages of discovery.

ISBN: 9781554890965   (paperback)

http://www.mmpubs.com/

Project Lessons from the Roman Empire: An Ancient Guide to Modern Project Management

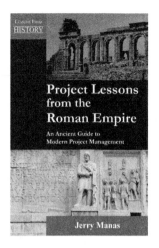

The leaders of the Roman Empire established many of the organizational governance practices that we follow today, in addition to remarkable feats of engineering using primitive tools that produced roads and bridges which are still being used today, complex irrigation systems, and even "flush toilets." Yet, the leaders were challenged with political intrigue, rebelling team members, and pressure from the competition. How could they achieve such long-lasting greatness in the face of these challenges?

Join author Jerry Manas as he takes you on a journey through history to learn about project management the Roman way. Discover the 23 key lessons that can be learned from the successes and failures of the Roman leadership, with specific advice on how they can be applied to today's projects.

Looking at today's hottest topics, from the importance of strategic alignment for your projects through to managing transformational change and fostering work/life balance while still maintaining overall performance, you'll find that the Romans already faced-and conquered-these challenges two thousand years ago. Read this intriguing book to learn how they did it.

ISBN: 9781554890545   (paperback)

http://www.mmpubs.com/

CPSIA information can be obtained
at www.ICGtesting.com
Printed in the USA
LVOW04s0417300416
486043LV00019BA/317/P